T0311654

DYNAMIC RISK ANALYSIS IN THE CHEMICAL AND PETROLEUM INDUSTRY

DYNAMIC RISK ANALYSIS IN THE CHEMICAL AND PETROLEUM INDUSTRY

Evolution and Interaction with Parallel Disciplines in the Perspective of Industrial Application

Edited by

NICOLA PALTRINIERI

FAISAL KHAN

Amsterdam • Boston • Heidelberg • London • New York • Oxford
Paris • San Diego • San Francisco • Singapore • Sydney • Tokyo

Butterworth-Heinemann is an imprint of Elsevier

Butterworth-Heinemann is an imprint of Elsevier
The Boulevard, Langford Lane, Kidlington, Oxford OX5 1GB, United Kingdom
50 Hampshire Street, 5th Floor, Cambridge, MA 02139, United States

Notices

Knowledge and best practice in this field are constantly changing. As new research and experience
broaden our understanding, changes in research methods, professional practices, or medical treatment
may become necessary.

Practitioners and researchers must always rely on their own experience and knowledge in evaluating
and using any information, methods, compounds, or experiments described herein. In using such
information or methods they should be mindful of their own safety and the safety of others, including
parties for whom they have a professional responsibility.

To the fullest extent of the law, neither the Publisher nor the authors, contributors, or editors,
assume any liability for any injury and/or damage to persons or property as a matter of products liability,
negligence or otherwise, or from any use or operation of any methods, products, instructions, or
ideas contained in the material herein.

Library of Congress Cataloging-in-Publication Data
A catalog record for this book is available from the Library of Congress

British Library Cataloguing-in-Publication Data
A catalogue record for this book is available from the British Library

ISBN: 978-0-12-803765-2

For information on all Butterworth-Heinemann publications
visit our website at https://www.elsevier.com/

Working together
to grow libraries in
developing countries

www.elsevier.com • www.bookaid.org

Publisher: Joe Hayton
Acquisition Editor: Fiona Geraghty
Editorial Project Manager: Maria Convey
Production Project Manager: Nicky Carter
Designer: Matthew Limbert

Typeset by TNQ Books and Journals

CONTENTS

CONTRIBUTORS

V. Casson Moreno
University of Bologna, Bologna, Italy

V. Cozzani
University of Bologna, Bologna, Italy

N.J. Edwin
Safetec, Trondheim, Norway; SINTEF Technology and Society, Trondheim, Norway

T.O. Grøtan
SINTEF Technology and Society, Trondheim, Norway

S. Hauge
SINTEF Technology and Society, Trondheim, Norway

N. Khakzad
Delft University of Technology, Delft, The Netherlands

F. Khan
Memorial University of Newfoundland, St John's, NL, Canada

K. Kyaw
Trondheim Business School, Norwegian University of Science and Technology (NTNU), Trondheim, Norway

G. Landucci
University of Pisa, Pisa, Italy

S. Massaiu
Institute for Energy Technology, OECD Halden Reactor Project, Halden, Norway

A. Matteini
University of Bologna, Bologna, Italy

W.R. Nelson
DNV GL, Houston, TX, United States

T. Østerlie
Norwegian University of Science and Technology (NTNU), Trondheim, Norway

N. Paltrinieri
Norwegian University of Science and Technology (NTNU), Trondheim, Norway; SINTEF Technology and Society, Trondheim, Norway

M. Pontiggia
D'Appolonia S.p.A., San Donato Milanese (MI), Italy

G. Reniers
University of Antwerp, Antwerp, Belgium; KU Leuven, Brussels, Belgium; Delft University of Technology, Delft, The Netherlands

E. Salzano
University of Bologna, Bologna, Italy

G.E. Scarponi
University of Bologna, Bologna, Italy

L. Talarico
University of Antwerp, Antwerp, Belgium

A. Tugnoli
University of Bologna, Bologna, Italy

V. Villa
University of Bologna, Bologna, Italy

H. Yu
University of Tasmania, Launceston, Australia

PREFACE

Static, incomplete, superficial, fictitious, or simply wrong. Application of risk analysis to process industries has been largely criticized and blamed in response to recent major accidents. Since it was first proposed, modifications and improvements have been made, and a generally accepted approach is considered in several regulations and standards. However, research in this field keeps producing new tools and techniques to overcome the flaws and deficiencies encountered in decades of use.

Risk analysis is about to enter an era of big data, where the main challenges are represented by the ability to provide continuous acquisition, effective process, and meaningful communication of such information. A new generation of wireless sensors, increasingly powerful computers, and optical fibers are only some examples of industry potential.

The focus of this book is the emerging topic of *dynamic risk analysis*, as opposed to traditional risk analysis, which is incapable of reflecting constantly evolving real-world risk. Dynamic risk analysis is often confused with online monitoring of representative indicators, but the adjective *dynamic* has a deeper significance, according to Merriam-Webster's, *(of a process or system) characterized by constant change, activity, or progress.*

Most risk analysis steps have the potential to be performed in a dynamic fashion. Inputs and models may be continuously improved, calibrated, and corrected based on new related evidence and lessons learned to progressively adapt to the ever-changing reality. A wider range of data may be considered: not only technical, but also human and organizational, information allowing for a proactive perspective. In addition, results may outline the economic impact of risk management to promote safety-oriented company strategies.

Although dynamic risk management is common practice in finance as a response to the financial crisis in 2008, in a safety-critical sector such as the chemical and petroleum industry, most methods for quantitative risk assessment mainly provide static evaluations. Standards (eg, ISO 31000 on risk management and NORSOK Z-013 on risk and emergency preparedness analysis) and relevant regulations (eg, the European Union's Seveso directives on the control of major-accident hazards involving dangerous substances) suggest updates of risk analysis, but mostly in conjunction with major changes in the plant or organization.

"Google Trends" shows that the number of Google searches for the term "big data" has increased about 100 times since 2011, and today it has reached its peak. On the contrary, the term "dynamic risk" had its peak in 2009 (after the financial crisis), and today its popularity on the search engine has decreased by about one third. Although many factors may affect such trends and do not represent the actual applications, this may reflect the challenges of dynamic risk analysis in finding its place in standard industrial approaches.

Research on how to dynamically assess risk has been carried out in several chemical and petroleum companies, but no real implementation has been attempted. Methodological evolution may be mirrored only partially by industrial application and is at risk to remain an unused intellectual exercise. The structure of the book addresses this issue by distinguishing between theoretical foundations and practical methodology tutorials. This allows for selective reading based on the reader's purpose. This book is not aimed to be an exhaustive review of dynamic risk analysis; it is rather a concrete support for the application of new risk analysis techniques.

Twenty concise chapters are grouped into four main parts, followed by a concluding chapter. Fig. 1 shows how they relate to each other:

1. introduction, providing an overview of the state of the art and new popular definitions
2. dynamic risk analysis, further divided into sections for each core step of risk analysis
3. interaction with parallel disciplines, transversal to some of risk analysis core steps
4. dynamic risk management, describing the overall management of dynamic methods

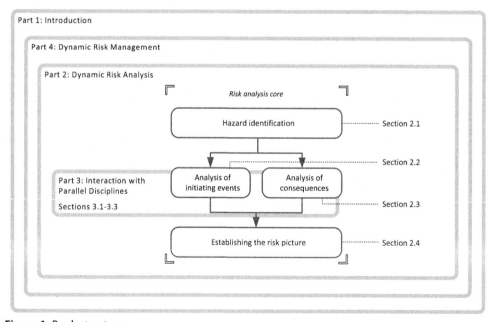

Figure 1 Book structure.

This volume is intended to become a support for (1) professionals, aiming to improve risk analysis by introducing innovative techniques and exploiting the potential of available technology, and (2) safety engineering students, who will be dealing with such approaches in the near future. Moreover, in the perspective of future research on this topic, this book aims to represent a concrete reference point within an ever-growing variety of innovations.

PART 1

Introduction

CHAPTER 1

A Short Overview of Risk Analysis Background and Recent Developments

V. Villa[1], N. Paltrinieri[2,3], F. Khan[4], V. Cozzani[1]

[1]University of Bologna, Bologna, Italy; [2]Norwegian University of Science and Technology (NTNU), Trondheim, Norway; [3]SINTEF Technology and Society, Trondheim, Norway; [4]Memorial University of Newfoundland, St John's, NL, Canada

1. INTRODUCTION

Public interest in the field of risk analysis has expanded greatly during the past three decades, and risk analysis has emerged as an essential and systematic tool that plays a relevant role in the overall management of many aspects of our lives. In technical domains characterized by risk of major accidents, risk assessment has dramatically shown its importance. For instance, despite the obvious differences between the nuclear and the chemical process sectors, a continuous exchange of knowledge and methods from one to the other has led to huge improvements in the chemical process industry [1] and helped it to cope with increasing issues of social acceptability [2]. Distant events in time, such as the tragedies that occurred in Bhopal (1984) and Piper Alpha (1988), as well as more recent ones, such as Buncefield (2005) and Deepwater Horizon (2010), have emphasized the essential role of adequate management and control for the chemical process industry.

The European industrial safety regulations aimed at controlling major-accident hazards related to chemical substances are named after the town of Seveso, Italy, the scene of a disaster in a chemical process plant in 1976. In 2012, the third generation of these regulations (Seveso III directive) [3] was issued; it applies to more than 10,000 industrial establishments in the European Union, mainly chemical, petrochemical, logistics, and metal refining sectors [4]. Quantitative risk assessment (QRA) is used to evaluate the overall process safety risk in the chemical process industry and identify areas requiring risk reduction [5], to comply with the related regulations.

In the past decades, several reviews dealt with risk analysis for process industries [6—9], but risk analysis methodologies and applications have rapidly evolved toward a dynamic direction, to address risk issues in a continuously evolving environment and to overcome the limitations of traditional techniques. This chapter summarizes risk analysis methodologies and relevant applications for a process industry, highlighting how recent techniques may overcome some of the drawbacks identified in conventional methods.

Dynamic Risk Analysis in the Chemical and Petroleum Industry
ISBN 978-0-12-803765-2

2. FUNDAMENTALS OF RISK ANALYSIS

2.1 Quantitative Risk Analysis

QRA is a systematic methodology for identifying and quantifying contributions to the overall risk of a process facility. As defined by NORSOK Standard Z-013 [10] and by the International Organization for Standardization/International Electrotechnical Commission standard [11], QRA includes establishment of context, risk identification, performance of risk analysis, and risk evaluation. Communication, consultation, monitoring, and review activities should be performed prior to, during, and after the assessment to guarantee the achievement of its goals. QRA can provide authorities and stakeholders with a sound basis for creating awareness about existing hazards and risks. Based on the outcomes from the QRA, potential measures to control or reduce risk can be implemented, and their effect can be assessed.

A preliminary step (Fig. 1.1, 1: establishing the context) defines objectives, responsibilities, and methods as well as risk acceptance criteria and deliveries throughout the

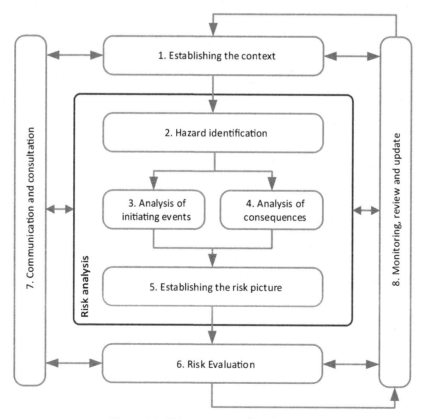

Figure 1.1 Risk assessment flowchart [10].

process and execution plan, to derive full value from the results obtained [12]. The following step in the development of a QRA [Fig. 1.1, 2: hazard identification (HAZID)] is the identification of hazards, which may have several important aims: it may highlight possible malfunctions of the systems, outline top events that are undesired situations, and describe potential scenarios associated with the top events and their consequences. As reported by the Center for Chemical Process Safety [5], several approaches to HAZID may be employed: checklist analysis, what–if analysis, preliminary hazard analysis, fault tree analysis, hazard and operability study, bow-tie analysis, etc. Their applicability depends on the project life cycle as well as the amount of information required. The maximum credible accident scenario analysis method developed by Khan and Abbasi [13] can be used as a criterion to identify credible scenarios among a large number of possibilities.

Estimation of potential accident frequencies and evaluation of event consequences are central steps for the whole QRA process (Fig. 1.1, 3: analysis of initiating events, and 4: analysis of consequences). Crowl and Louvar [14] state that risk analysis basically involves the estimation of accident frequencies and consequences using engineering and mathematical techniques. One way to retrieve generic failure frequencies and probabilities is to use databases and apply the information in QRA calculations; specific plant data should be applied, if available. Guidelines in the QRA "Purple Book" [15] report generic loss of containment events and failure frequencies for a number of standard installations such as storage tanks, transport units, pipelines, and loading equipment. Consequence estimation is used to determine the potential for damage or injury from specific unwanted events. Quantification of consequences has usually been carried out in terms of losses in production, human health, assets, and environment. The assessment of consequences can be performed using a number of physical-mathematical and empirical models. A description of many available approaches has been presented by Arunraj and Maiti [16]. A review of available damage models applied to calculate the spatial distribution of damage (eg, probability of human death) has been carried out by Cozzani and Salzano [17].

The risk picture is established by considering all the risk contributions of the scenarios analyzed for a specific area of the plant, which, in turn, are obtained through the composition of the related frequencies and damages (Fig. 1.1, 5: establishing the risk picture). The choice of risk metrics is critical because it directs what kind of information to obtain from the risk analysis. Johansen and Rausand provide a short review of such risk metrics [18]. Risk is finally evaluated against the acceptance criteria previously defined (Fig. 1.1, 6: risk evaluation), and, if needed, risk mitigation measures are taken. Two steps are continuously performed along the whole QRA process: "communication and consultation" (Fig. 1.1, 7) and "monitoring, review and update" (Fig. 1.1, 8). In fact, it must be underlined that QRA should be considered an iterative procedure and should be updated at the maximum every 5 years or in case of major plant changes, as stated by the Seveso directive [3].

2.2 Applications, Accomplishments, and Limitations of QRA in the Chemical Process Industry

At the beginning of its history, QRA was used primarily as a verification activity [19], whereas now it plays a relevant role in different aspects of the process plant life cycle. An important application of QRA is risk-based design; as pointed out by Fadier and De la Garza [20] and Hale et al. [21], risk-based or risk-informed design plays a relevant role in risk reduction for chemical and process installations. A valid quantitative approach to a risk-based engineering design involves acquiring and incorporating all possible knowledge about the design into the decision process [22]. For instance, safety system modeling is now an integral part of risk assessment studies and represents a significant application of QRA. In this regard, safety systems aiming to avoid, prevent, limit, or control accidents are evaluated to examine the extent to which they are effective in reducing the risk of an accident. During the operation of chemical plants, a risk-based approach to maintenance takes on a major role, determining the risk associated with failure modes and allowing a pertinent maintenance strategy to develop [23]. QRA has also proved its effectiveness in its application to an entire industrial area [24] and in several subsequent developments with land use planning purposes [25,26]. Moreover, in the past years, much attention has been placed on the integration of QRA with quantitative assessment of cascading events. A summary of the mentioned approaches can be found in Cozzani et al. [27]. Security issues, such as possible terrorist attacks and sabotage [28], are increasingly included in QRAs to define an exhaustive risk picture that represents the real situation in a detailed and realistic way. Greenberg et al. [29] describe some of the latest accomplishments that generic risk analysis has achieved in the past years addressing health, safety, and the environment. Accomplishments can be identified for QRA along the same lines as what was identified for generic risk analysis. On the other hand, there are frequent criticisms associated with QRA paving the way for further improvements. A summary of risk analysis accomplishments and limitations has been reported in Table 1.1.

3. WAY FORWARD: DYNAMIC RISK ASSESSMENT APPROACHES

Conventional risk analysis methodologies suffer the disadvantage of being intrinsically static, and this may preclude possible updates and integrations with the overall risk figure (Table 1.1). For this reason, during the past decade, several efforts have been devoted to the development of dynamic assessment and management approaches considering the evolution of conditions affecting risk.

Some of the first attempts to simulate the dynamic nature of system behavior were made by Swaminathan and Smidts, who proposed a methodology to extend the application of event sequence diagrams to the modeling of dynamic situations and identification of missing accidental scenarios [33]. Čepin and Mavko developed an extension of the

Table 1.1 Accomplishments and Limitations in Conventional Risk Analysis for the Chemical Process Industry [29]

Accomplishments	Limitations
• Risk perception and behavior are increasingly affected by analytical evaluation. Risk managers may be perceived to be more competent by means of QRA, which leads to less concern and more benefits perceived by the public and greater acceptance of a hazard. • QRA has laid the foundation for estimating the economic impact of hazardous events. Several examples show that such impact should not be disregarded and represent an important element for critical decision-making. • By means of QRA, risk communication is providing reliable and useful information to all interested parties, including scientists and managers "who too often think that they already know whatever they need to know." • Legal decisions may be now supported by the application of QRA for a more risk-informed outcome, as demonstrated by the outcome of the trials assessing the responsibilities of the explosion in the Buncefield oil depot in 2005 [30].	• Creedy [31] addresses the estimation of the frequencies included in the QRA by stating, "it still appears to be largely based on values from several decades ago." There is a need for realistic values for both failure probability and event frequency that better describe the actual conditions of the system. • Classic risk analysis approach is static. It decomposes a system and focuses on failure events of components. This approach is not sufficient to explain all that can go wrong, because it does not grasp the dynamics of unsafe interactions and fails to capture the variation of risks as deviations or changes in the process and plant [5]. • Apostolakis [32] claims that probability cannot be realistically calculated, meaning that one cannot use straightforward statistical methods and divide the number of failures by the number of trials to calculate "realistic" probabilities. QRA analysts make extensive use of expert judgment and should always look at the uncertainties associated with the results.

fault tree analysis to represent time requirements in safety systems [34]. Similarly, Bucci et al. [35] presented a methodology to extend fault trees and event trees in a dynamic perspective.

The first complete dynamic risk assessment methodology for process facilities, named dynamic failure assessment, was developed by Meel and Seider [36]. This approach aims to estimate the dynamic probabilities of accident sequences, including near misses and incident data (named accident sequence precursors) as well as real-time data from processes.

Kalantarnia et al. [37] integrated Bayesian failure mechanisms with consequence assessment. Starting from this foundational contribution, several methodologies have tried to improve the approach by introducing slight modifications. For instance, hierarchical Bayesian analysis widened the field of application for dynamic risk assessment to rare events, owing to a two-stage Bayesian method [38]. System hazard identification, prediction, and prevention methodology is another derived approach specifically addressing accident modeling, which integrates technical and nontechnical barriers [39].

Another mentionable contribution is the dynamic operational risk assessment methodology [40], which included conceptual framework design, mathematical modeling, and decision-making based on cost—benefit analysis. Bayesian inference risk assessment was also coupled with a dynamic HAZID technique named DyPASI (dynamic procedure for atypical scenarios identification) [41]. The DyPASI procedure is based on conventional bow-tie analysis and allows the HAZID process to systematically update on the basis of relevant early warnings and risk notions [42].

The Center for Integrated Operations in the Petroleum Industry has recently developed the "risk barometer" technique [43], which aims to continuously monitor risk picture changes and support decision makers in daily operations. This proactive approach to risk is based on the availability of a large amount of real-time data, for which collection is made easier by the extensive use of information and communication technologies.

3.1 Potential Improvements and Limitations

Dynamic risk assessment may be potentially applied not only in the design stage of a process but also throughout its lifetime, allowing safer operations and easier maintenance, as well as supporting precise, risk-informed, and robust decision-making. As reported in Table 1.2, dynamic approaches may show advantages in design and operation of processing facilities, whereas the main limitations are shared between design and operation phases.

Dynamic methods may improve design and comparison between different alternatives of safety systems by integrating standard processes of HAZID with notions on emerging risks or external experience with relevant events. Accounting for human and organizational factors since the beginning of the design process would also allow for proactive prevention of potential underlying issues. The operation phase of a plant would be the main phase to benefit from dynamic risk assessment. In fact, frequent reiteration of risk assessment may allow continuous evaluation of safety measures, refinement of their management, enhanced management of safety-critical operations, and improved maintenance planning. Moreover, constant monitoring of human and organizational factors would allow for feedback on the organization's safety culture and support training sessions focusing on key organizational issues.

The next challenges for dynamic risk assessment will be devoted to solving the issues of formalization, standardization, and creation of automated software able to perform dynamic risk assessment. Although several steps forward have been made from the pioneering studies, every dynamic approach may be effective only if associated with a proper safety culture that continuously searches for learning opportunities by monitoring and recording process performance and incidents [44]. Moreover, dynamic methods may be further developed to include both external (eg, natural hazards) and internal cascading events (eg, domino effects).

Table 1.2 Potential Improvements and Limitations of Dynamic Risk Assessment

	Potential Improvements	Limitations
Design	Potential improvements affecting design and management of safety systems: • Dynamic methods for hazard identification may improve development of scenario generation and description by means of integration with emerging risks, early warnings, and external experience with relevant scenarios. • Integration with technical, human, and organizational factors from the beginning of the design process may improve frequency and consequence evaluation. • Comparison between different safety measure alternatives may be improved by risk assessment refinement and evaluation of related costs and benefits.	Limitations transversal to design and operation: • No regulations are currently available for dynamic risk assessment methodologies and applications. • Effectiveness of the methods relies on continuous monitoring activity and real-time data capturing. It implies the necessity to collect early warnings, near misses, incidents, and accident data. • Most of the methodologies presented herein are part of ongoing studies. There are still some issues to be addressed for each method (eg, for Bayesian inference techniques, the use of free-distribution data [41]; for Bayesian networks, the net development [45]; for risk barometers, the indicators aggregation processes [46]). • No automated software is available, and there is very limited experience in industry. • Conventional models are needed as prerequirement (eg, bow-tie for DyPASI, conventional QRA for risk barometer [47]). • Dynamic methods may need further development, in particular regarding inclusion of cascading events (eg, natural hazards, domino effects).
Operation	Potential improvements affecting safety-critical operations: • Risk assessment may be reiterated more frequently in comparison with traditional methods. • Additional safety measures may be evaluated during the operational phase, based on updated scenarios, to fulfill risk minimization. • Management of critical operations may be improved to prevent accidents by reviewing operating conditions and determining whether to proceed. • Lacking/defective maintenance may be effectively detected; this would optimize inspection and maintenance time intervals. • Human and organizational issues may be detected during the operation phase to improve general safety culture.	

4. CONCLUSIONS

In the past 30 years, risk analysis techniques have strengthened process safety and showed their usefulness in supporting the process industry business by enabling risk management. Despite the obvious fact that it is not an exact description of reality, QRA proved to be an effective analytic predictive tool to assess the risks of complex chemical process systems. However, this short overview pointed out that better refinement of risk assessment tools is required to achieve its full potential of applicability.

Most of the research is oriented toward improving basic aspects, such as data frequencies. The main development pathway is given by the application of dynamic approaches as a direct consequence of the now-feasible possibility of real-time monitoring for process facilities. Dynamic risk assessment aims to take into account new risk notions and early warnings, and to systematically update the related risk, ensuring enhanced flexibility. In particular, the dynamic risk assessment approach, coupled with a dynamic procedure for HAZID, seems to be a promising step forward. Another valuable development of dynamic risk assessment is represented by a novel proactive method based on indicators: the risk barometer.

To conclude, despite the fact that risk assessment application has greatly increased the safety of chemical process plants, major-accident scenarios are still occurring. To guarantee a certain level of safety, it was shown that risk assessment techniques should be constantly improved and should evolve in parallel with the increasing complexity of the systems where they are applied.

REFERENCES

[1] Charvet C, Chambon J-L, Corenwinder F, Taveau J. Learning from the application of nuclear probabilistic safety assessment to the chemical industry. Journal of Loss Prevention in the Process Industries 2011;24:242—8. http://dx.doi.org/10.1016/j.jlp.2010.09.007.

[2] Marshall VC. The social acceptability of the chemical and process industries: a proposal for an integrated approach. Chemical Engineering Research and Design 1997;75:145—55. http://dx.doi.org/10.1016/S0263-8762(97)80012-3.

[3] EU. SEVESO III. Directive 2012/18/EU of the European Parliament and of the Council of 4 July 2012 on the control of major-accident hazards involving dangerous substances, amending and subsequently repealing Council Directive 96/82/EC. 2012.

[4] European Commission — Environment Directorate. The Seveso Directive— Prevention, preparedness and response. Eur Comm Website 2015, http://ec.europa.eu/environment/seveso/.

[5] CCPS — Center for Chemical Process Safety. Guidelines for chemical process quantitative risk analysis. New York, USA: American Institute of Chemical Engineers (AIChE); 2000.

[6] Khan F, Abbasi SA. Techniques and methodologies for risk analysis in chemical process industries. Journal of Loss Prevention in the Process Industries 1998;11:261—77. http://dx.doi.org/10.1016/S0950-4230(97)00051-X.

[7] Tixier J, Dusserre G, Salvi O, Gaston D. Review of 62 risk analysis methodologies of industrial plants. Journal of Loss Prevention in the Process Industries 2002;15:291—303. http://dx.doi.org/10.1016/S0950-4230(02)00008-6.

[8] Marhavilas PK, Koulouriotis D, Gemeni V. Risk analysis and assessment methodologies in the work sites: on a review, classification and comparative study of the scientific literature of the period 2000—2009. Journal of Loss Prevention in the Process Industries 2011;24:477—523. http://dx.doi.org/10.1016/j.jlp.2011.03.004.

[9] Necci A, Cozzani V, Spadoni G, Khan F. Assessment of domino effect: State of the art and research needs. Reliability Engineering and System Safety 2015;143:3—18. http://dx.doi.org/10.1016/j.ress.2015.05.017.

[10] NORSOK. Standard Z-013, risk and emergency preparedness analysis. 3rd ed. Lysaker, Norway: Standards Norway; 2010.

[11] ISO31000:2009. Risk management: principles and guidelines. Geneva, Switzerland. 2009.

[12] Mannan MS. Lees's loss prevention in the process industries. 3rd ed. Burlington, MA, USA: Elsevier Butterworth-Heinemann; 2005.

[13] Khan F, Abbasi SA. A criterion for developing credible accident scenarios for risk assessment. Journal of Loss Prevention in the Process Industries 2002;15:467—75. http://dx.doi.org/10.1016/S0950-4230(02)00050-5.

[14] Crowl DA, Louvar JF. Chemical process safety: fundamentals with applications. 3rd ed. Boston, MA, USA: Prentice Hall; 2011.

[15] TNO. The "Purple Book" — Guidelines for quantitative risk assessment. CPR 18 E. Publ. Ser. Danger. Subst. (PGS 3). 2005.

[16] Arunraj NS, Maiti J. A methodology for overall consequence modeling in chemical industry. Journal of Hazardous Materials 2009;169:556—74. http://dx.doi.org/10.1016/j.jhazmat.2009.03.133.

[17] Cozzani V, Salzano E. The quantitative assessment of domino effects caused by overpressure. Part I. Probit models. Journal of Hazardous Materials 2004;107:67—80. http://dx.doi.org/10.1016/j.jhazmat.2003.09.013.

[18] Johansen IL, Rausand M. Risk metrics: interpretation and choice. IEEE International Conference on Industrial Engineering and Engineering Management 2012:1914—8. http://dx.doi.org/10.1109/IEEM.2012.6838079.

[19] Falck A, Skramstad E, Berg M. Use of QRA for decision support in the design of an offshore oil production installation. Journal of Hazardous Materials 2000;71:179—92. http://dx.doi.org/10.1016/S0304-3894(99)00078-3.

[20] Fadier E, De la Garza C. Safety design: towards a new philosophy. Safety Science 2006;44:55—73. http://dx.doi.org/10.1016/j.ssci.2005.09.008.

[21] Hale AR, Kirwan B, Kjellén U. Safe by design: where are we now? Safety Science 2007;45:305—27. http://dx.doi.org/10.1016/j.ssci.2006.08.007.

[22] Demichela M, Piccinini N. Risk-based design of a regenerative thermal oxidizer. Industrial and Engineering Chemistry Research 2004;43:5838—45. http://dx.doi.org/10.1021/ie0342208.

[23] Arunraj NS, Maiti J. Risk-based maintenance-techniques and applications. Journal of Hazardous Materials 2007;142:653—61. http://dx.doi.org/10.1016/j.jhazmat.2006.06.069.

[24] Egidi D, Foraboschi FP, Spadoni G, Amendola A. The ARIPAR project: analysis of the major accident risks connected with industrial and transportation activities in the Ravenna area. Reliability Engineering and System Safety 1995;49:75—89. http://dx.doi.org/10.1016/0951-8320(95)00026-X.

[25] Spadoni G, Egidi D, Contini S. Through ARIPAR-GIS the quantified area risk analysis supports land-use planning activities. Journal of Hazardous Materials 2000;71:423—37. http://dx.doi.org/10.1016/S0304-3894(99)00091-6.

[26] Spadoni G, Contini S, Uguccioni G. The new version of ARIPAR and the benefits given in assessing and managing major risks in industrialised areas. Process Safety and Environmental Protection: Transactions of Institution of Chemical Engineers Part B 2003;81:19—30. http://dx.doi.org/10.1205/095758203762851958.

[27] Cozzani V, Antonioni G, Landucci G, Tugnoli A, Bonvicini S, Spadoni G. Quantitative assessment of domino and NaTech scenarios in complex industrial areas. Journal of Loss Prevention in the Process Industries 2014;28:10—22. http://dx.doi.org/10.1016/j.jlp.2013.07.009.

[28] Reniers G, Audenaert A. Preparing for major terrorist attacks against chemical clusters: intelligently planning protection measures w.r.t. domino effects. Process Safety and Environmental Protection 2013;92:583—9. http://dx.doi.org/10.1016/j.psep.2013.04.002.

[29] Greenberg M, Haas C, Cox A, Lowrie K, McComas K, North W. Ten most important accomplishments in risk analysis, 1980—2010. Risk Analysis 2012;32:771—81. http://dx.doi.org/10.1111/j.1539-6924.2012.01817.x.

[30] HSE — Health and Safety Executive. Buncefield: Why Did It Happen? 2011. p. 36. www.hse.gov.uk/comah/buncefield/buncefield-report.pdf.

[31] Creedy GD. Quantitative risk assessment: How realistic are those frequency assumptions? Journal of Loss Prevention in the Process Industries 2011;24:203—7. http://dx.doi.org/10.1016/j.jlp.2010.08.013.

[32] Apostolakis GE. How useful is quantitative risk assessment? Risk Analysis 2004;24:515—20. http://dx.doi.org/10.1111/j.0272-4332.2004.00455.x.

[33] Swaminathan S, Smidts C. Identification of missing scenarios in ESDs using probabilistic dynamics. Reliability Engineering and System Safety 1999;66:275—9. http://dx.doi.org/10.1016/S0951-8320(99)00024-1.

[34] Čepin M, Mavko B. A dynamic fault tree. Reliability Engineering and System Safety 2002;75:83—91. http://dx.doi.org/10.1016/S0951-8320(01)00121-1.

[35] Bucci P, Kirschenbaum J, Mangan LA, Aldemir T, Smith C, Wood T. Construction of event-tree/fault-tree models from a Markov approach to dynamic system reliability. Reliability Engineering and System Safety 2008;93:1616—27. http://dx.doi.org/10.1016/j.ress.2008.01.008.

[36] Meel A, Seider W. Real-time risk analysis of safety systems. Computers and Chemical Engineering 2008;32:827—40. http://dx.doi.org/10.1016/j.compchemeng.2007.03.006.

[37] Kalantarnia M, Khan F, Hawboldt K. Modelling of BP Texas City refinery accident using dynamic risk assessment approach. Process Safety and Environmental Protection 2010;88:191—9. http://dx.doi.org/10.1016/j.psep.2010.01.004.

[38] Khakzad N, Khan F, Paltrinieri N. On the application of near accident data to risk analysis of major accidents. Reliability Engineering and System Safety 2014;126:116—25. http://dx.doi.org/10.1016/j.ress.2014.01.015.

[39] Rathnayaka S, Khan F, Amyotte P. SHIPP methodology: predictive accident modeling approach. Part I: methodology and model description. Process Safety and Environmental Protection 2011;89:151—64. http://dx.doi.org/10.1016/j.psep.2011.01.002.

[40] Yang X, Mannan MS. The development and application of dynamic operational risk assessment in oil/gas and chemical process industry. Reliability Engineering and System Safety 2010;95:806—15. http://dx.doi.org/10.1016/j.ress.2010.03.002.

[41] Paltrinieri N, Khan F, Cozzani V. Coupling of advanced techniques for dynamic risk management. Journal of Risk Research 2014:1—21. http://dx.doi.org/10.1080/13669877.2014.919515.

[42] Paltrinieri N, Tugnoli A, Buston J, Wardman M, Cozzani V. Dynamic procedure for atypical scenarios identification (DyPASI): A new systematic HAZID tool. Journal of Loss Prevention in the Process Industries 2013;26:683—95. http://dx.doi.org/10.1016/j.jlp.2013.01.006.

[43] Hauge S, Okstad E, Paltrinieri N, Edwin N, Vatn J, Bodsberg L. In: Handbook for monitoring of barrier status and associated risk in the operational phase, the risk barometer approach. Trondheim, Norway: SINTEF F27045; 2015.

[44] Paltrinieri N, Khan F, Amyotte P, Cozzani V. Dynamic approach to risk management: application to the Hoeganaes metal dust accidents. Process Safety and Environmental Protection 2014;92:669—79. http://dx.doi.org/10.1016/j.psep.2013.11.008.

[45] Pasman HJ, Rogers WJ. Bayesian networks make LOPA more effective, QRA more transparent and flexible, and thus safety more definable! Journal of Loss Prevention in the Process Industries 2013;26:434—42. http://dx.doi.org/10.1016/j.jlp.2012.07.016.

[46] Paltrinieri N, Hokstad P. Dynamic risk assessment: development of a basic structure. In: Saf. Reliab. Methodol. Appl. — Proc. Eur. Saf. Reliab. Conf. Esrel 2014, Wroclaw, Poland; 2015. p. 1385—92. http://dx.doi.org/10.1201/b17399-191.

[47] Villa V, Paltrinieri N, Cozzani V. Overview on dynamic approaches to risk management in process facilities. Chemical Engineering Transactions 2015;43:2497—502. http://dx.doi.org/10.3303/CET1543417.

CHAPTER 2

New Definitions of Old Issues and Need for Continuous Improvement

N. Paltrinieri[1,2], F. Khan[3]
[1]Norwegian University of Science and Technology (NTNU), Trondheim, Norway; [2]SINTEF Technology and Society, Trondheim, Norway; [3]Memorial University of Newfoundland, St John's, NL, Canada

1. INTRODUCTION

Major industrial accidents can be grouped under several definitions on the basis of their specific features. Particular attention has been recently focused on low-probability, high-impact events, which are challenging not only to prevent but also to grasp. In fact, they may be the result of incomplete hazard identification, poor knowledge management, or ineffective likelihood evaluation, or they may be simply unpredictable. This chapter reports a brief overview of definitions of such extreme events, highlighting and discussing similarities, differences, and limitations.

2. ATYPICAL ACCIDENT SCENARIOS

Paltrinieri et al. [1,2] define an accident scenario as atypical when it is "not captured by hazard identification methodologies because deviating from normal expectations of unwanted events or worst case reference scenarios." An atypical accident may occur when hazard identification does not produce a complete overview of system hazards. This step represents a qualitative preassessment and is the foundation of the whole risk management process. Paltrinieri et al. [1,3] attribute the potential deficiencies of hazard identification results to lack of specific knowledge and low awareness of related risks by analysts and, more in general, safety professionals in an organization. In particular, risk awareness has a primary role, as illustrated in Fig. 2.1.

Fig. 2.1 shows the development of two different cases, both starting from a condition of unawareness of accident risk and lack of related information (point 1). In an ideal case, a reasonable doubt would grow in the minds of safety professionals of an organization once they come across some examples of early warnings, such as unwanted events occurring in their site, historical data, and literature studies from external sources. Thus, their risk awareness would progressively increase with the availability of related information (point 2), which is used to increase the knowledge of the potential scenario to a condition of relative confidence about accident risk (point 3), where only loss of organizational memory may reduce awareness. According to the Table 2.1 definitions, the ideal case

Dynamic Risk Analysis in the Chemical and Petroleum Industry
ISBN 978-0-12-803765-2

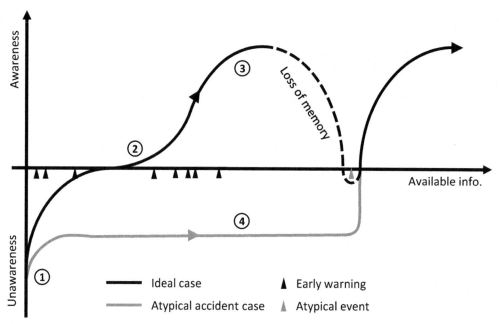

Figure 2.1 Management of accident risk on the basis of awareness and availability of related information in an ideal case and an atypical accident case. *(Adapted from Paltrinieri N, Dechy N, Salzano E, Wardman M, Cozzani V. Lessons learned from Toulouse and Buncefield disasters: from risk analysis failures to the identification of atypical scenarios through a better knowledge management. Risk Analysis 2012;32:1404−19.)*

Table 2.1 Definitions of Known/Unknown Events (Used by US Army Intelligence and Made Popular by Donald Rumsfeld [4])

Unknown knowns	Known knowns
Events we are not aware that we (can) know by means of available (but disregarded) information	Events we are aware that we know, for which risk can be managed with a certain level of confidence
Unknown unknowns	**Known unknowns**
Events we are not aware that we do not know, for which risk cannot be managed	Events we are aware that we do not know, for which we employ both prevention and learning capabilities

depicts proactive management of accident risk, where the accident is initially an "unknown unknown" (point 1) that first evolves into a "known unknown" (point 2) and is finally and successfully classified as a "known known" (point 3).

In the "atypical accident" case, available information on risk accident is disregarded. The situation develops in a condition of unawareness despite the succession of related early warnings (point 4), and the only development is the passage from the risk of "unknown unknowns" to "unknown knowns" that may be potentially understood through the available information (Table 2.1). It is only when the atypical accident occurs that risk awareness suddenly increases and compensation measures are taken.

The literature shows several examples of atypical accidents that occurred or may have the potential to occur (Table 2.2). The former are cases of hazard identification failure: The respective safety reports identified relatively less critical accident scenarios, despite

Table 2.2 Examples of Atypical Accident

Atypical event	Past Accidents		Potential Accidents	
Atypical event	Toulouse, France, 2001 Ammonium nitrate explosion	Vapor cloud explosion, Buncefield, UK, 2005	Potential accident with carbon capture and sequestration	Potential accident with liquefied natural gas regasification
Explanation	Worst case scenario in safety report: ammonium nitrate storage fire	Worst case scenario in safety report: large pool fire	Increased scale of substance handling and relative lack of experience in the identification of related hazards	
Early warnings	Previous similar accidents: • Oppau, Germany, 1921 • Texas City, US, 1947 • Brest, France, 1947 • Red Sea, cargo ship Tirrenia, 1954	Previous similar accidents: • Houston, US, 1962 • Baytown, US, 1977 • Newark, US, 1983 • Naples, Italy, 1985 • St. Herblain, France, 1991 • Jacksonville, USA, 1993 • Laem Chabang, Thailand, 1999	Examples of past related accidents: • Natural CO_2 releases, lakes Monoun and Nyos, Cameroon, 1984 and 1986 • Cold boiling liquid expanding vapor explosion, Worms, Germany, 1988	Examples of past related accidents: • Rapid phase transition, Canvey Island, UK, 1973 • Asphyxiation, Oklahoma City, US, 1978 • Rapid phase transition, Bontang, Borneo, 1993 • Boiling liquid expanding vapor explosion, Tivissa, Spain, 2002
References of atypical event study	[1,5,6]	[1,5,6]	[7—10]	[11,12]

the occurrence of previous similar events in the past. The latter are related to new and emerging technologies: Related hazards may be well known to specialists but still a gray area for safety professionals, who nonetheless have the opportunity to learn from early warnings.

3. BLACK SWANS

Until the 17th century, all swans known by Europeans were white. However, with the discovery of Australia, the first black swans were sighted, which became the symbol of disproved belief (ie, all swans are white). For this reason, a black swan event is a rare event that has never been encountered before and can be summarized by the following three principles [13]:

- Rarity: a black swan is an outsider because nothing could convincingly point to its possibility.
- Impact: a black swan has extreme consequences.
- Predictability: a black swan can be explained only after the fact and cannot be anticipated.

These principles may lead to the degradation of predictability and the need for robustness against negative black swans and for learning from positive ones. In fact, they are not only adverse events but also rare unplanned opportunities from which we should learn. Taleb [13] affirms that these kinds of events are the result of epistemic limitations (or distortions), demonstrating a clear overlapping with the definition of atypical events [1]. In fact, Aven and Aven and Krohn [14,15] refer to a black swan as a surprisingly extreme event relative to one's belief/knowledge, defining and giving examples of three types of black swans (partially according to the concept of unknowns/knowns previously introduced; see Table 2.1): unknown unknowns, unknown knowns, and events judged negligible.

An example of a black swan of the unknown unknown type is represented by the swine flu spread in 2009 caused by the H1N1 virus, for which a vaccine was quickly developed. In some countries the authorities aimed at vaccinating the entire population. The influenza turned out to be relatively mild; however, the vaccination had severe side effects, which were previously unknown [16]. A black swan of the unknown known type might have been the disaster involving the Deepwater Horizon drilling rig, where a worker did not alert others on the rig as pressure increased in the drilling pipe, a sign of possible gas/fluid entry into the wellbore (kick), which can lead to blowout [17]. The Fukushima nuclear disaster, because of its low probability of occurrence, was judged as a negligible event. This third type of black swan was preceded in past centuries by extreme natural events (tsunamis of heights beyond the design criterion of the nuclear plant), which were not accounted for during the design of the nuclear reactors [18].

Black swans are addressed by Paté-Cornell [18] in comparison with another definition of rare event: the perfect storm. This kind of event involves mostly aleatory uncertainties (randomness) in conjunction with rare but known events. It takes its name from the devastating storm in the northern Atlantic described by Junger [17], which caught some boats by surprise and killed 12 people in 1991. It was the result of the conjunction of a storm from the North American mainland, a cold front from the north, and a tropical storm from the south. Paté-Cornell affirms that black swans represent the ultimate epistemic uncertainty or lack of fundamental knowledge, where not only the distribution of a parameter but also, in the extreme, the very existence of the phenomenon itself is unknown. However, in reality, most scenarios involve both aleatory and epistemic uncertainties, and a clear distinction cannot be outlined.

4. DRAGON KINGS

Like black swans, the events defined by Sornette [19] as dragon kings are outliers. The term "king" emphasizes the importance of such extreme events, which may be identified as the outliers in a power law distribution. This is an analogy to the wealth of kings, which, if plotted against their subjects' wealth, is beyond the power law distribution. These exceptional events are also defined as "dragons" to stress that they are beyond the normal, with extraordinary characteristics; the presence of these, if confirmed, has profound significance.

The definition of dragon kings also refers to the existence of a transient organization into extreme events that are statistically different from the rest of their smaller siblings. This realization opens the way for a systematic theory of predictability of catastrophes, contrasting with the definition of black swans. The concept is rooted in geophysics [20]. One of Sornette's earlier works was related to the prediction of earthquakes. He saw that some degrees of organization and coordination could serve to amplify fractures, which are always present and forming in the tectonic plates. Organization and coordination may turn small causes into large effects, ie, explosive ruptures such as earthquakes, which are characterized by low probability. This physical model suggests the possibility of predicting such events. In fact, if time between these smaller fracture events decreases with a specific log-periodic pattern, earthquake probabilities are much higher.

Dragon kings may represent an answer to Paté-Cornell [18] and Haugen and Vinnem [21], who warn against the misuse of the concept of the black swan as a reason for ignoring potential scenarios or waiting until a disaster happens to take safety measures and issue regulations against a predictable situation.

An example of dragon king was searched for among the accidents recorded in the MHIDAS (major hazard incident data service) database [22], which collected about 1500 events from 1916 to 1992. The accidents were ranked on the basis of the number of fatalities they caused and plotted in a double logarithmic ranking/fatalities diagram

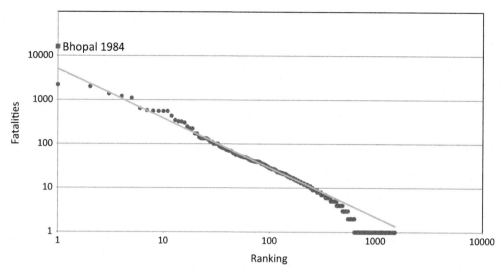

Figure 2.2 Rank-ordering plot of the fatalities caused by major hazard incident data service accidents [22].

(Fig. 2.2). The distribution of the events follows the power law with regularity, as shown by their good approximation to the straight line in Fig. 2.2. However, the recorded accident beyond the power law, causing the highest number of fatalities, may be the disaster that occurred in Bhopal in 1984, if it is considered to have caused all 16,000 fatalities claimed [23]. (Other sources report the number of fatalities to be around 3400 [24].)

If the Bhopal disaster can be defined as a dragon king, then an emerging organization of small previous "fractures" into the system of the chemical plant should be recognized. Several authors have identified a "list of things that went wrong simultaneously at Bhopal" [24–28]:

- Many safety-related devices were not well designed.
- The plant was losing money, which resulted in staff and maintenance budget cutbacks.
- There was a social system that dismissed safety culture and created extreme tension between management and workers.
- The plant was to close permanently, which affected operator morale and contributed to the lack of maintenance and the bypassing of safety systems.
- There was complete failure or lack of an emergency response program.
- There was ineffective treatment of the injured.
- There were no citations or any awareness of the dangers of this plant.

These fractures led to an extraordinary amplification of the event, turning it into a large-scale effect. Their coordination could potentially have been identified, however, and such an accident predicted, as theorized by Sornette.

5. SMALL THINGS

Despite the different shades of their definitions, most of the authors cited in this chapter agree on one point: extreme accidents are the result of a combination of events, which often cannot be classified as extreme or unpredictable. In fact, such causal events may range from repetitive technical failures to common human errors.

This is demonstrated by Paltrinieri et al. [1] in the identification of the chain of events that led to the atypical accident in Buncefield in 2005 (see Table 2.2). For instance, one cause was the frequent oil tank overfilling events as a result of operators often not informing others about the changes in the incoming fuel flow rate [29]. Although Taleb [13] points at the consequentiality of black swans, Aven [30] reports the events and conditions that combined to lead to the Deepwater Horizon accident, previously mentioned as an example of a black swan. Paté-Cornell [18], in describing perfect storms, refers to a "conjunction of known phenomena," which is "an alignment of factors whose possibilities are known separately but whose simultaneity seems extremely unlikely." Finally, Sornette [19] bases his theory on the aggregation of relatively smaller events leading to extreme consequences, because "in the theory of complex systems and in statistical physics big disruptions do not need large perturbations to occur." As complexity of a system increases, it enhances system susceptibility means, and minute changes would have disproportional consequences.

Preventing small things from happening might be the key to breaking this chain of events, because there are always potential accident scenarios that we are not aware we do not know—unknown unknowns. Awareness of their likelihood should always be present, and appropriate precautions to consciously face our relative ignorance (shifting from prevention of unknown unknowns to the prevention of known unknowns) would probably lower the probability of extreme events. Small things might be represented by recurring old issues in a plant or an organization, which do not need imaginative definitions to be prevented but perhaps only compliance with already present procedures. Thus, on one hand, we should avoid excessive focus on uncertainties in risk analysis that may have the side effect of disregarding our consolidated knowledge. On the other hand, new definitions to attribute to major accidents are the expression of the need for continuous improvement and adaptation with increasingly complex systems, such as today's chemical process industry. Emerging technologies, tighter interdependencies with critical infrastructures, and increasing volumes of handled hazardous substances may potentially hide unexperienced risks. Management of such risks requires high levels of risk awareness, opening to dynamic updating, and capability of reorienteering industrial activities toward intrinsically safer conditions. This approach may be described in a few words with "learning from experience," a relatively small thing (conceptually less complex) with dramatically heighted implications.

6. CONCLUSIONS

Low-probability, high-impact events may feature different attributes. Atypical accidents are events that were disregarded in the hazard identification phase owing to a combined lack of awareness and knowledge, underlined by the classification into unknown unknowns and unknown knowns. Although black swans are completely aleatory, dragon kings generate hope regarding the predictability of their formation and occurrence. However, all these classifications of major accidents agree on the relative complexity of their development. In fact, conjunctions of small things and tangled cause-consequence relationships among them have the potential to result in extreme effects. For this reason, a twofold approach should be adopted to limit major accidents. Well-known small failures, the multiplicity of which creates a fertile ground for disasters, should not be disregarded but seriously and comprehensively addressed. At the same time, continuous improvement of the models and classifications employed to keep track of industry evolution and increasing complexity is essential. Such a dynamic approach should be adopted at different levels and include ongoing improvement and integration of risk assessment techniques to progressively approximate real-world conditions.

REFERENCES

[1] Paltrinieri N, Dechy N, Salzano E, Wardman M, Cozzani V. Lessons learned from Toulouse and Buncefield disasters: from risk analysis failures to the identification of atypical scenarios through a better knowledge management. Risk Analysis 2012;32:1404–19.
[2] Paltrinieri N, Dechy N, Salzano E, Wardman M, Cozzani V. Towards a new approach for the identification of atypical accident scenarios. Journal of Risk Research 2013;16:337–54.
[3] Paltrinieri N, Oien K, Cozzani V. Assessment and comparison of two early warning indicator methods in the perspective of prevention of atypical accident scenarios. Reliability Engineering & System Safety 2012;108:21–31.
[4] Rumsfeld D. In: News transcript. Washington, DC, USA: U.S. Department of Defense, Office of the Assistant Secretary of Defense (Public Affairs); February 12, 2002.
[5] Paltrinieri N, Cozzani V, Wardman M, Dechy N, Salzano E. Atypical major hazard scenarios and their inclusion in risk analysis and safety assessments. Reliability, Risk and Safety—Back to the Future 2010:588–95.
[6] Paltrinieri N, Dechy N, Salzano E, Wardman M, Cozzani V. Towards a new approach for the identification of atypical accident scenarios. Journal of Risk Research 2012;16:337–54.
[7] Paltrinieri N, Wilday J, Wardman M, Cozzani V. Surface installations intended for Carbon Capture and Sequestration: atypical accident scenarios and their identification. Process Safety and Environmental Protection 2014;92:93–107.
[8] Wilday J, Paltrinieri N, Farret R, Hebrard J, Breedveld L. Addressing emerging risks using carbon capture and storage as an example. Process Safety and Environmental Protection 2011;89:463–71.
[9] Wilday J, Paltrinieri N, Farret R, Hebrard J, Breedveld L. Carbon Capture and storage: a case study of emerging risk issues in the iNTeg-Risk project. In: Institution of chemical Engineers Symposium Series. 156th ed. 2011. p. 339–46.
[10] Paltrinieri N, Breedveld L, Wilday J, Cozzani V. Identification of hazards and environmental impact assessment for an integrated approach to emerging risks of CO_2 capture installations. Energy Procedia 2013:2811–8.

[11] Paltrinieri N, Tugnoli A, Cozzani V. Hazard identification for innovative LNG regasification technologies. Reliability Engineering & System Safety 2015;137:18–28.

[12] Paltrinieri N, Tugnoli A, Bonvicini S, Cozzani V. Atypical scenarios identification by the DyPASI procedure: application to LNG. Chemical Engineering Transactions 2011:1171–6.

[13] Taleb N. The black swan: the impact of the highly improbable. New York: Random House; 2007.

[14] Aven T. On the meaning of a black swan in a risk context. Safety Science 2013;57:44–51.

[15] Aven T, Krohn BS. A new perspective on how to understand, assess and manage risk and the unforeseen. Reliability Engineering & System Safety 2014;121:1–10.

[16] Munsterhjelm-Ahumada K. Health authorities now admit severe side effects of vaccination swine flu pandemrix and narcolepsy. Orthomolecular Medicine News Releases 2012.

[17] Financial post. Deepwater rig worker weeps as he admits he overlooked warning of blast that set off America's worst environmental disaster. 2013. http://business.financialpost.com/2013/03/14/halliburton-worker-weeps-as-he-admits-he-overlooked-warning-of-blast-that-set-off-americas-biggest-oil-spill-in-gulf/?__lsa=42e0-28bb.

[18] Paté-Cornell E. On "black swans" and "perfect storms": risk analysis and management when statistics are not enough. Risk Analysis 2012;32:1823–33.

[19] Sornette D. Dragon-kings, black swans and the prediction of crises. ETH Zurich, Chair of System Design 2009.

[20] Musgrave GL, Weatherall JO. The physics of wall street: a brief history of predicting the unpredictable. Business Economics 2013;48:203–4.

[21] Haugen S, Vinnem JE. Perspectives on risk and the unforeseen. Reliability Engineering & System Safety 2015;137:1–5.

[22] MHIDAS (Major Hazard Incident Data Service). Mhidas database. Harwell, UK: AEA Technology, Major Hazards Assessment Unit, Health and Safety Executive; 2003.

[23] Eckerman I. The Bhopal saga: causes and consequences of the world's largest industrial disaster. Universities Press; 2005.

[24] Labib A. Learning (and unlearning) from failures: 30 years on from Bhopal to Fukushima an analysis through reliability engineering techniques. Process Safety and Environmental Protection 2015;97: 80–90.

[25] Khan FI, Abbasi SA. Major accidents in process industries and an analysis of causes and consequences. Journal of Loss Prevention in the Process Industries 1999;12:361–78.

[26] Yang M, Khan F, Amyotte P. Operational risk assessment: a case of the Bhopal disaster. Process Safety and Environmental Protection 2015;97:70–9.

[27] Chouhan TR. The unfolding of Bhopal disaster. Journal of Loss Prevention in the Process Industries 2005;18:205–8.

[28] Eckerman I. The Bhopal gas leak: analyses of causes and consequences by three different models. Journal of Loss Prevention in the Process Industries 2005;18:213–7.

[29] Health and Safety Executive. Buncefield: Why did it happen? Bootle, UK: HSE; 2011.

[30] Aven T. Comments to the short communication by Jan Erik Vinnem and Stein Haugen titled "perspectives on risk and the unforeseen". Reliability Engineering & System Safety 2015;137:69–75.

PART 2

Dynamic Risk Analysis

SECTION 2.1

Hazard Identification

CHAPTER 3

Advanced Technique for Dynamic Hazard Identification

N. Paltrinieri[1,2], A. Tugnoli[3], V. Cozzani[3]

[1]Norwegian University of Science and Technology (NTNU), Trondheim, Norway; [2]SINTEF Technology and Society, Trondheim, Norway; [3]University of Bologna, Bologna, Italy

1. INTRODUCTION

In the past decades European directives and international standards [1–5] pressed industry toward the development and extended use of structured hazard identification techniques. In particular, the "Seveso" directives [1–3], concerning the control of major-accident hazards involving dangerous substances, require the issuing of "safety reports" when identification and assessment of potential accident scenarios is needed [6]. However, despite such requirements and the measures taken, completely unexpected and unmitigated events still occur in the European chemical process industry [7–12], highlighting the need for improvement of current hazard identification techniques. The main issue is represented by the availability of methods able to identify low-probability accidents within a routine hazard identification process, capturing evidence of new hazards and learning from early warnings once they come to light. For this reason a specific method named dynamic procedure for atypical scenarios identification (DyPASI) was developed to continuously improve the process of hazard identification.

2. A FRAMEWORK FOR THE IDENTIFICATION OF ATYPICAL SCENARIOS

As stated in Chapter 2, an accident scenario is defined here as "atypical" if it is not captured by conventional hazard identification techniques because it deviates from normal expectations of unwanted events or worst-case reference scenarios [12]. This definition may include accident scenarios well known to specialists but not to the community of safety professionals, because their occurrence may be very rare or limited to date to specific industrial activities [13].

Two significant examples of atypical accident scenarios that occurred in the European Union are those that took place in Toulouse and Buncefield, respectively, in 2001 and 2005. The explosion at the "off-specifications" ammonium nitrate warehouse of the nitrogen fertilizer factory AZF (Grande Paroisse) in Toulouse caused 30 fatalities and €1.5 billion in damages. However, the worst scenario considered by the safety report

Dynamic Risk Analysis in the Chemical and Petroleum Industry
ISBN 978-0-12-803765-2

for the site was a storage fire [11]. At the oil depot of Buncefield, a vapor cloud explosion caused £1 billion of damage but fortunately no fatalities [9]. In this case the worst scenario considered in the hazard identification process was a low-severity gasoline pool fire [7].

These two accidents had been anticipated by several similar past accidents: Many severe ammonium nitrate explosions occurred between 60 and 90 years ago, and vapor cloud explosions involving gasoline and light hydrocarbon fuels have occurred in oil depots on average every 5 years since mid-1960 [14]. Furthermore, after 2005 two other similar vapor cloud explosions took place within a few days of each other: in Bayamón (Puerto Rico) on October 23, 2009, [15] and in Jaipur (India) on October 29, 2009 [16]. This highlights that early warnings are not always considered, and insights from past accidents may not be considered in daily activities concerning hazard identification.

A different type of latent risk can be represented by the accident scenarios related to new and emerging technologies. When new technologies are implemented, the potential for specific accident scenarios may still need to be properly identified and may remain unidentified until an accident or a near miss takes place for the first time. Examples of new and emerging technologies can be found within the fields of liquefied natural gas regasification (see Chapter 4) and carbon capture and storage [17]. New and alternative technologies are implemented for these processes, and the scale and extent of substance handling is set to increase dramatically worldwide. A lack of substantial operational experience may lead to difficulties in identifying accurately the hazards associated with the process. For this reason, such new and emerging hazards may comply with the definition of atypical scenarios previously discussed.

Hazard identification is the first step of the risk assessment process [18,19], and in this preliminary activity the hazards are usually identified only qualitatively. Risk quantification pertains to the later steps of conventional techniques, being out of the scope of the hazard identification process [18]. Nevertheless, the hazard identification process is an important part of risk management because no action can be made to avoid, or mitigate, the risk deriving from unidentified hazards. The hazard identification process also has a large potential for human error, with little or no feedback pertaining to those errors. The aforementioned atypical accidents are the severe feedback of such errors.

3. STATE OF THE ART

Several extensive reviews of available hazard identification techniques are present in literature [19–22]. Table 3.1 reports representative hazard identification methods for each phase of a plant life cycle, with related benefits and limitations. Some methods are relatively easy to apply, but hazards can be missed owing to limitations in the experience of the analysts (what-if analysis) or in the method itself (checklists). In the case of more structured methods, such as preliminary hazard analysis, fault tree analysis, and hazard and operability study (HAZOP), the more systematic they are, the better the coverage of

Table 3.1 Examples of Hazard Identification Methods for Each Phase of a Plant Life Cycle [18,21]

Plant Life-Cycle Phase	Example of Method	Description	Benefits (+) and Limitations (−)
Feasibility	What-if analysis	"What-if" analysis [18,19] uses a creative brainstorming methodology. The assessment asks a number of questions that begin with "what if" to attempt to identify any associated hazards. The results are often documented in a table format and will often contain questions, related hazards or consequences, safeguards present at the time, and recommendations, if applicable.	+ Easy to apply − Experienced assessors required or hazards can be missed − Time consuming for complex processes
Concept	Preliminary hazard analysis	Preliminary hazard analysis [23–25] is used as an early means of hazard identification. It follows an approach similar to HAZOP, although it splits the process into larger sections, generally major process items and associated lines and heat exchangers.	+ Facilitates the building of fault trees and event trees + Systematically identifies the accident scenarios + Easy to perform + Aids in the production of a more inherently safe process − Will not identify all the causes − Will only identify and examine the major hazards
Detailed design & engineering/installation	Fault tree analysis	Fault tree analysis [18] is a graphical representation of the combination of faults leading to a predefined undesired event. The methodology uses logic gates to show all credible paths from which the undesired event could occur. The fault tree is developed from the top down (ie, from the undesired top event to the primary events that initiated the failure), and the logic gates (AND/OR gates) indicate the passage of the fault logic up the tree.	+ Able to produce quantitative results + Shows a logical representation of the sequence of events + Can be used to assess a wide range of failures − Time consuming and expensive for complex systems − Experienced assessment team required − Some top events might be missed

Continued

Table 3.1 Examples of Hazard Identification Methods for Each Phase of a Plant Life Cycle [18,21]—cont'd Plant Life-Cycle

Phase	Example of Method	Description	Benefits (+) and Limitations (−)
Operation	HAZOP	A hazard and operability study [26] can be used at varying times during the life cycle of the process, including hazard assessment of modifications. It produces a comprehensive evaluation of the process through a number of guidewords (typically no/none, more, less, part of, reverse, other than, as well as) combined with parameters (flow, pressure, temperature, reaction, level, composition) and is systematically applied to each pipe and vessel of the process.	+ Systematic and comprehensive technique + Examines the consequences of the failure − Time consuming and expensive − Requires detailed design documentation to perform the full study − Additional guidewords required for unusual hazard − Requires experienced practitioners − Focuses mostly on single-event causes of deviation
Decommissioning	Checklist analysis	Checklists [18] produce a detailed examination of the process plant by applying experience of everyday operations and previous incidents in similar plants. Only a coarse list should be used to aid in the direction of the work to stimulate the inventiveness of the team members. Once the brainstorming has been completed, the detailed checklist can then be used to identify areas that have been overlooked.	+ Easy to apply + Can be performed by inexperienced practitioners − Assessment only as complete as the list used − Not easy to apply to novel decommissioning processes

potential accident scenarios. However, this leads to more time consuming and expensive applications, progressively requiring more and more experienced analysts.

Theoretical and practical limitations affecting results of hazard identification suggest the need for an improvement of current techniques. In particular, the Center for Chemical and Process Safety [18] identified five main hazard identification principles whose potential deficiencies should be taken into account for the creation of novel methods:

1. Completeness: some accident situations, causes, and effects may have been unintentionally neglected;
2. Reproducibility: assumptions from the analysts may affect results;
3. Inscrutability: results may be difficult to understand and synthesize for use;
4. Relevance of experience: lack of specific experience may lead one to neglect the significance of some aspects; and
5. Subjectivity: analysts may have to use their judgment when extrapolating from their experience.

4. DYNAMIC PROCEDURE FOR ATYPICAL SCENARIOS IDENTIFICATION

DyPASI was developed to provide comprehensive hazard identification including atypical scenarios. DyPASI is a hazard identification method aiming at the systematization of information from early signals of risk related to past accident events, near misses, and literature studies. It supports the identification and the assessment of atypical potential accident scenarios related to the substances, the equipment, and the industrial site considered. DyPASI is one of the results of the European Commission FP7 iNTeg-Risk project [27], which addressed the management of emerging risks.

The application of DyPASI entails a systematic screening process that based on early warnings and risk notions should be able to identify possible atypical scenarios available at the time of the analysis. The well-established approach of bow–tie analysis [28], which aims to identify all the potential major accident scenarios that may occur in an industrial site, is taken as a basis to develop the methodology. Specific branches may be integrated consistently with the bow–tie diagrams and related safety barriers defined for the newly identified scenarios (Fig. 3.1). Further details on how to apply DyPASI are reported in Chapter 4.

Table 3.1 lists well-known hazard identification techniques and associated life-cycle phases of the plant where their application is more appropriate. DyPASI, however, may be suitable for application in each phase of the plant life cycle. In particular, the application is also possible in the early phases of research and development and of conceptual design. This feature, which is not present in most conventional hazard identification techniques, may be exploited to focus notions about the hazards embedded in the process and to orient efforts toward safer solutions [29]. As such, DyPASI may contribute to

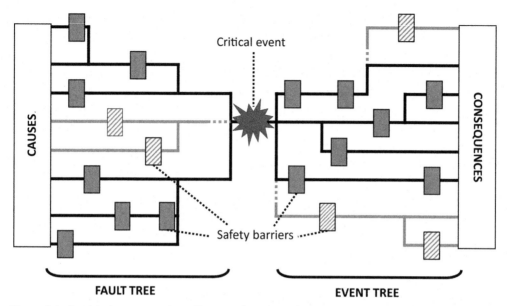

Figure 3.1 Graphical representation of bow-tie diagram with integration of branches newly identified through dynamic procedure for atypical scenarios identification.

reducing the unavailability of early hazard identification tools, which is a well-known issue in inherent safety implementation [30].

DyPASI is a tool specifically defined for the continuous improvement of risk management, providing a procedure to enable systematic updating of the hazards identified and managed in the process. DyPASI may either be used as a stand-alone technique or be coupled with existing conventional techniques. In the latter case, it may effectively integrate the existing hazard identification methods to obtain more exhaustive results. In particular it provides a structured and yet dynamic approach in the retrieval of information from early warnings and atypical scenarios. The format of the results from DyPASI allows for integration with the hazard identification techniques based on fault tree and event tree analysis, effectively extending the applicability of DyPASI from preliminary hazard analysis to the detailed assessment of complex systems.

4.1 Hazard Identification Deficiencies Tackled by DyPASI

The DyPASI technique is based on the recommendations drawn by Paltrinieri et al. [12] from the analysis of lessons learned from atypical accident scenarios. The five principles in the application of hazard identification techniques identified by the Center for Chemical and Process Safety [18], listed previously, were also pursued to develop the technique.

4.1.1 Completeness

The main purpose of DyPASI is to achieve a more comprehensive overview of the potential hazards in a plant by means of a systematic screening of related early warnings.

To this end, DyPASI can be dynamically applied when there is evidence of new experiences or a change in safety notions. The application requires the definition of the information need and of the search boundaries with the purpose of reducing the information overload from database search. However, it is the prioritization of information that helps the user to identify the early warnings by means of an ordered classification of data, focused on relevance and impact. The integration of the atypical scenarios identified into the bow-tie diagram is carried out to provide results that stick to the conventional tools used in process hazard identifications and that may be more easily integrated with those coming from other techniques.

4.1.2 Reproducibility

DyPASI helps prevent the practice of "jumping to conclusions" (described by Kahneman in "Thinking Fast and Slow" [31]), which would allow a savings of time and effort if the conclusions are correct but with costs in the case of a mistake that are not acceptable for the hazard identification process. Following this technique step by step, as explained in Chapter 4, would reduce unclear assumptions from the analyst and would help one to better ponder the accident scenarios to consider and their early warnings. On the other hand, this technique specifically aims to obtain new results every time it is reiterated, with the purpose to dynamically learn from experience and take effective actions of precaution and control. Thus, the identification of new scenarios should not be affected by analyst assumptions, but rather it should be the result of a transparent process of search and learning of related early warnings.

4.1.3 Inscrutability

To avoid inscrutability of the hazard identification results, the bow-tie technique was used as a basis in developing DyPASI because it "in its visual form makes the analysis easy to understand, and can show what safeguards protect against particular initiating causes and loss event consequences" [18]. Moreover, as Chapter 4 shows, DyPASI borrows concepts from the set theory with the purpose of systematically and consistently integrating general patterns of atypical scenarios into prior bow-tie diagrams. Following Chapter 4 guidelines is not a vain logic exercise but aims to obtain clear and concise results and avoid harmful repetitions resulting from the blind addition of new diagram elements.

4.1.4 Relevance of Experience

The process of information retrieval explicitly deals with lack of experience in hazard identification. This search for learning opportunities is supported by a clear definition of information needs and search boundaries. General principles used to define automated information retrieval systems, such as web engines, are also used to define an appropriate approach to the information access process. Relevance judgment is suggested beginning with the first information results, so as to tune the search query for more pertinent data. The pieces of information are systematized in an ordered register, where relevance and impact are highlighted. Thus, even if this is a qualitative prioritization and it is a task of the user to extrapolate early warnings, the study and classification of information aims to help him or her obtain a more transparent process and avoid unjustified neglect.

4.1.5 Subjectivity

As with all qualitative hazard identification techniques, DyPASI ultimately relies on the user's judgment. However, this systematic approach not only would make the user further ponder his or her judgments and justify actions but would also allow easy integration and updating of the results whenever any lacks or weaknesses are spotted. The entire risk assessment process depends on having performed an exhaustive search for what can go wrong [18]. Using DyPASI does not completely cancel a user's related subjectivity but aims to make it more transparent and to make the whole process more flexible to later adjustments.

4.2 Limitations and Integration With Other Techniques

The possibility of capturing atypical scenarios during the initial hazard identification phase heavily depends on the experience of the user. The DyPASI method was built to systematically approach the issue of critical event identification, screening, and organizing available information. The systematic approach of DyPASI should limit the possibility of failing in the identification of some relevant accident scenario, so as to give the analyst a chance to obtain or update comprehensive results.

DyPASI should not be considered as a stand-alone technique but should be integrated with other risk management tools for a more complete approach to the problem. In fact, it can be easily adapted and applied to various bow-tie methodologies whenever enhancement or updating of hazard identification analysis is needed. DyPASI can be easily coupled with a technique for the development of early warning indicators to further revamp a proactive approach in the prevention of major accidents. As proposed by Paltrinieri [32], this would be an effective strategy in which organizational and technical factors are addressed by an integrated approach.

The use of this integrated approach for the prevention of atypical events is suggested in the case of new processes, new technologies, and change in social or public perception as well as when new scientific knowledge allows a long-standing issue to be identified as a risk.

5. CONCLUSIONS

The DyPASI methodology was built in an effort to mitigate deficiencies of the current hazard identification techniques. The technique represents a translation of the recommendations and lessons outlined from past atypical accidents. The main aim of the methodology is to provide easier but comprehensive hazard identification for the industrial process analyzed. The principal features of DyPASI are its dynamic and systematic nature and the enhancement of the knowledge management on which the hazard identification process is based. Its limitations are mainly related to the fact that DyPASI cannot directly address underlying human and organizational factors. DyPASI is a tool to support emerging risk management process, helping to trigger a gradual process of identification and assimilation of previously unrecognized atypical scenarios.

REFERENCES

[1] Council Directive 82/501/EEC of 24 June 1982. On the major-accident hazards of certain industrial activities. Official Journal of the European Communities 1982;L 230(1).
[2] Council Directive 96/82/EC of 9 December 1996. On the control of major-accident hazards involving dangerous substances. Official Journal of the European Communities 1996;L 010(13).
[3] Directive 2012/18/EU of the European Parliament and of the Council of 4 July 2012. On the control of major-accident hazards involving dangerous substances, amending and subsequently repealing Council Directive 96/82/EC Text with EEA relevance. Official Journal of the European Communities 2012;L 197(1).
[4] ISO. Guidelines on tools and techniques for hazard identification and risk assessment. ISO 17776: 2000. Geneva, Switzerland: ISO; 2010.
[5] NORSOK. Risk and emergency preparedness assessment. Z-013. Oslo, Norway: Standards Norway; 2010.
[6] Wood MH, Fabbri L, Struckl M. Writing Seveso II safety reports: new EU guidance reflecting 5 years' experience with the Directive. Journal of Hazardous Materials 2008;157:230−6.
[7] Atkinson G, Buston J, Salzano E, Dechy N, Van Wijk L, Joyce B, et al. Atypical vapour cloud explosion type events, deliverable D1.4.4.2. Buxton, UK. 2010.
[8] Atkinson G, Coldrick S, Gant S, Cusco L. Flammable vapor cloud generation from overfilling tanks: learning the lessons from Buncefield. Journal of Loss Prevention in the Process Industries 2015;35: 329−38.
[9] Buncefield Major Incident Investigation Board The Buncefield Incident 11 December 2005. The final report of the major incident investigation board. Bootle, UK: HSE Books; 2008.
[10] Dechy N, Bourdeaux T, Ayrault N, Kordek M-A, Le Coze J-C. First lessons of the Toulouse ammonium nitrate disaster, 21st September 2001, AZF plant, France. Journal of Hazardous Materials 2004; 111:131−8.
[11] Dechy N, Mouilleau Y. Damages of the Toulouse disaster, 21st September 2001. In: Proceedings of the 11th International Symposium loss prevention 2004; 2004. Praha, 31 May − 3 June 2004.
[12] Paltrinieri N, Dechy N, Salzano E, Wardman M, Cozzani V. Lessons learned from Toulouse and Buncefield disasters: from risk analysis failures to the identification of atypical scenarios through a better knowledge management. Risk Analysis 2012;32:1404−19.
[13] Kletz T. Will cold petrol explode in the open air? Loss Prevention Bulletin 2006;188:9.
[14] Paltrinieri N, Dechy N, Salzano E, Wardman M, Cozzani V. Towards a new approach for the identification of atypical accident scenarios. Journal of Risk Research 2012;16:337−54.
[15] Chemical Safety Board. Caribbean petroleum refinery tank explosion and fire. 2009.
[16] Committee Govt of India. Independent Inquiry Committee Report on Indian Oil Terminal Fire at Jaipur on 29th October 2009. 2010.
[17] Paltrinieri N, Wilday J, Wardman M, Cozzani V. Surface installations intended for Carbon Capture and Sequestration: atypical accident scenarios and their identification. Process Safety and Environmental Protection 2014;92:93−107.
[18] Center for Chemical Process Safety. Guidelines for chemical process quantitative risk analysis. New York, USA: America Institute of Chemical Engineers (AIChE); 2000.
[19] Lees FP. Loss prevention in the process industries: hazard identification, assessment and control. Butterworths; 1980.

[20] Crawley F, Tyler B, Centre EPS. Hazard identification methods: institution of chemical engineers. 2003.

[21] Glossop M, Ioannides A, Gould J. Review of hazard identification techniques. 2005. Bottle, UK.

[22] Khan FI, Abbasi SA. Techniques and methodologies for risk analysis in chemical process industries. Journal of Loss Prevention in the Process Industries 1998;11:261–77.

[23] Wells G, Wardman M, Whetton C. Preliminary safety analysis. Journal of Loss Prevention in the Process Industries 1993;6:47–60.

[24] Wells G. Hazard identification and risk assessment. Rugby, Warwickshire: Institution of Chemical E, IChemE; 1997.

[25] Ozag H, Bendixen LM. Hazard identification and quantification. Chemical Engineering Progress 1987;83:55–64.

[26] Kletz TA. Hazop and hazan. Institution of Chemical Engineers; 2001.

[27] Paltrinieri N, Tugnoli A, Buston J, Wardman M, Cozzani V. Dynamic procedure for atypical scenarios identification (DyPASI): a new systematic HAZID tool. Journal of Loss Prevention in the Process Industries 2013;26:683–95.

[28] Delvosalle C, Fievez C, Pipart A, Debray B. ARAMIS project: a comprehensive methodology for the identification of reference accident scenarios in process industries. Journal of Hazardous Materials 2006;130:200–19.

[29] Kletz TA, Amyotte P. Process plants: a handbook for inherently safer design. 2nd ed. CRC Press; 2010.

[30] Khan FI, Amyotte PR. How to make inherent safety practice a reality. Canadian Journal of Chemical Engineering 2003;81:2–16.

[31] Kahneman D. Thinking, fast and slow: Farrar, Straus and Giroux; 2011.

[32] Paltrinieri N, Øien K, Cozzani V. Assessment and comparison of two early warning indicator methods in the perspective of prevention of atypical accident scenarios. Reliability Engineering and System Safety 2012;108:21–31.

CHAPTER 4

Dynamic Hazard Identification: Tutorial and Examples

N. Paltrinieri[1,2], A. Tugnoli[3], V. Cozzani[3]

[1]Norwegian University of Science and Technology (NTNU), Trondheim, Norway; [2]SINTEF Technology and Society, Trondheim, Norway; [3]University of Bologna, Bologna, Italy

1. INTRODUCTION

The dynamic procedure for atypical scenarios identification (DyPASI) was developed to allow the systematization of information from early signals of risk related to past incident events, near misses, and inherent studies based on a structured review approach carried out in the hazard identification process. The methodology aims to complement and organize available knowledge and to define and take into account rare potential accident scenarios related to the substances, the equipment, and the industrial process under consideration. In this chapter, the DyPASI methodology is applied for hazard identification of new and alternative technologies for liquefied natural gas (LNG) regasification.

2. METHODOLOGY TUTORIAL

The flowchart in Fig. 4.1 illustrates the steps of the DyPASI technique for hazard identification.

2.1 Step 0: Preliminary Activity

As a preliminary activity, DyPASI requires the application of the conventional bow-tie technique to identify the relevant critical events. It should be recalled that the bow-tie technique combines fault tree and event tree analysis, which are merged to share a common element called critical event (see Fig. 4.3). The development of bow-ties can be performed following conventional guidelines such as those outlined by the Center for Chemical Process Safety [1]. As an alternative, the MIMAH tool (methodology for the identification of major accident hazards) can be applied [2].

2.2 Step 1: Retrieval of Risk Notions

A search for relevant information concerning undetected potential hazards and accident scenarios that may not have been considered in conventional bow-tie development is carried out. Typology of information needed and examples of search engines to be used in this search are reported in Table 4.1.

Dynamic Risk Analysis in the Chemical and Petroleum Industry
ISBN 978-0-12-803765-2

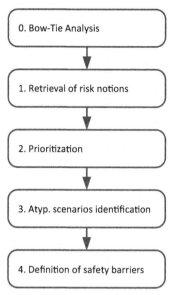

Figure 4.1 DyPASI flowchart.

To obtain good relevance of the results, search boundaries must be outlined and quoted in the formulation of the query, in the combination and number the analyst considers most appropriate. Examples of search boundaries used in queries are the site, the process, the equipment, the substance, and the substance state. Algorithms can be applied to rank the relevance of the results obtained. They compare the entries of the query with the information in the retrieved results. An example of a similarity ranking algorithm is proposed by Zhao et al. [4].

Table 4.1 Information Need and Examples of Search Systems for the DyPASI Information Retrieval Step [3]

Information Typology	Examples of Search Systems
• Past events • Past deviations from normal conditions • Experimental tests • Theoretical studies related to potential hazards, health, and environmental and societal consequences	• ARIA (French Ministry of Environment) • FACTS (TNO [Netherlands]) • Google Scholar (Google Inc.) • Internal company databases • MARS (JRC [European Union]) • MHIDAS (HSE [United Kingdom]) • SciVerse Scopus (Elsevier) • The Accident Database (IChemE, Institution of Chemical Engineers) • Web of Science (Thomson Reuters)

Table 4.2 Accident Scenario Severity Levels

Severity Level	Description
Near miss	An event that does not result in an actual loss but that has the potential to do so.
Mishap	An event that could cause minor health effects and/or minor impact to property and the environment.
Incident	An event that could cause major health effect or injury, localized damage to assets and environment, considerable loss of production, and damage to one's reputation.
Accident	An event that may cause one or more fatalities or permanent major disabilities, and/or heavy financial loss.
Disaster	An event that could cause multiple fatalities and extensive damage to property, system, and production. It may cause a shutdown of the plant for a significant time period, sometimes forever.

2.3 Step 2: Prioritization

Once the necessary information is gathered, a determination is made whether the data are significant enough to trigger further action and whether to proceed with the process of risk assessment. To support the process of prioritization, a register that collects the risk notions obtained from the retrieval process and shows their relative relevance and impact can be obtained. Possible consequences can be determined on the basis of the risk notions. The actual or potential final consequences of the relevant scenarios can be ranked by defining a severity level for the related scenarios (Table 4.2).

Definition of the potential impact of atypical scenarios can become guesswork. For this reason, Table 4.3 was outlined to suggest basic correlations between properties of the substance handled, potential consequences, and worst-case scenario severity.

It will be a task of the user of DyPASI to extrapolate early warnings of a potential atypical accident scenario. The classification of gathered data performed on the basis of relevance and impact will help the user to identify the most pertinent and serious signals.

2.4 Step 3: Atypical Scenarios Identification

Potential scenarios are isolated from the early warnings gathered, and a cause-consequence chain consistent with the bow-tie diagram is developed. At first, the initial cause, the critical event, and the final consequence, which are the main elements of a bow-tie diagram, must be identified within the potential atypical scenario. Then the other elements are defined asking the question "why," or, more specifically, "what is directly necessary and sufficient to cause this event," starting from the final consequence and going backward through the critical event to the initial cause. Once a specific pattern to describe the atypical scenario is defined, one or more suitable bow-tie diagrams obtained in step 0 may be identified for the process of integration. The specific search

Table 4.3 Potential Scenario Impact on the Basis of Substance Properties

	Properties of the Substance				
Consequences	Toxic	Oxidizing	Explosive/ Explosive dust	Flammable	Dangerous for Environment
Pool fire				Accident	
Tank fire				Accident	
Jet fire				Accident	
Vapor cloud explosion				Disaster	
Flash fire				Accident	
Toxic cloud	Disaster		Incident	Incident	
Fire		Accident	Accident	Accident	
Missile ejection		Accident	Accident	Accident	
Blast wave		Disaster	Disaster	Incident	
Fireball				Accident	
Major accident to environment	Incident		Mishap	Incident	Incident
Boilover				Accident	

boundaries used in the previous steps and the defined critical event should be used for the identification of appropriate bow-tie diagrams. If no suitable diagram is identified, the atypical scenario pattern must be considered a new bow-tie diagram itself, which should be added to the set of hazard identification results. The integration of the atypical scenario pattern should be performed considering half the diagram at a time and should move level by level from the critical event to the initial cause (if the fault tree section is considered) or final consequence (if the event tree is considered). The detailed mathematical model is described elsewhere [3].

2.5 Step 4: Definition of Safety Barriers

Past experience concerning the effectiveness and performance of safety barriers may be encompassed in the analysis. The integrated bow-tie diagrams including the atypical scenarios should be completed considering safety barriers, classified by their effectiveness and their function, as listed in the following:

- *To avoid*: safety function acting upstream of the bow-tie diagram event aiming to suppress the inherent conditions that cause it
- *To prevent*: safety function acting upstream of the bow-tie diagram event aiming to reduce its occurrence

- *To control*: safety function acting upstream of the fault tree event in response to a drift, which may lead to the event and safety function acting downstream of the event tree event aiming to stop it
- *To limit*: safety function acting downstream of the bow-tie diagram event aiming to mitigate it

Safety barriers can be physical and engineered systems or human actions based on specific procedures, or administrative controls that can directly implement the safety functions described (Fig. 4.2) [2].

To encompass past experience, safety barriers that acted properly at the moment of past accidents should be marked in green. Green is also applied to effective safety barriers in the case of near misses. Safety barriers that showed deficiencies in at least one past accident are marked in orange. New potential and hopefully more effective safety barriers are represented using red. This activity can also provide important elements for the risk mitigation process in the decision-making phase of risk management.

2.6 Follow-up

Follow-up consists of carrying out conventional risk management. Relevance of the risk related to the identified atypical scenarios should be assessed. Judgment of risk acceptability and the possible implementation of further specific safety barriers (or system modification) is demanded at this phase, which falls out of the scope of the DyPASI methodology.

DyPASI also requires a reiteration of the procedure, on the basis of the following criteria:

- after changes in process, equipment, workplace, or organization, to keep track of potential risk variation of the new system
- after changes in social or public perception, to assess long-standing issues newly considered as risk
- after new evidence in scientific knowledge allowing the identification of emerging issues
- at given time intervals (tentatively 2–5 years, depending on the system features), to prevent any of the above factors from going undetected

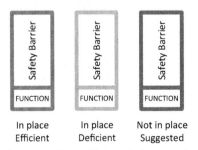

Figure 4.2 Safety barrier classification, where efficient/deficient refers to the counteraction in past accidents.

Table 4.4 Risk Notions Collected Through the Application of DyPASI [5,11,12]

Risk notion	Site	Process	Equipment	Substance	State	References
Evidence of the possibility of rapid phase transition reported in studies and standards	/	/	/	Natural gas	Liquefied	[13–15]
Rapid phase transition event following accidental leak (Canvey Island [UK], 1973)	Import terminal	Unloading	/	Natural gas	Liquefied	[16]
Rapid phase transition event following accidental leak (Bontang [Indonesia], 1993)	Export facility	Transport	Pipe	Natural gas	Liquefied	[17]
Boiling liquid expanding vapor explosion event of a liquefied natural gas road tanker − 1 fatality (Tivissa [Spain], 2002)	Road tanker	Transport	Tank	Natural gas	Liquefied	[18]
Several studies on liquefied natural gas risk considering terrorist attack	Facilities, shipyards, vessels, pipelines, fields.	/	/	Natural gas	Gaseous and liquefied	[19–21]
Cryogenic burns following accidental leak − 1 fatality (Arzew [Algeria], 1977)	Liquefied natural gas export facility	Ship loading	Large-diameter valve	Natural gas	Liquefied	[17,22]
Study addressing the issue of liquefied natural gas cryogenic damage	/	/	/	Natural gas	Liquefied	[13]
Asphyxiation event − 4 fatalities (Oklahoma City [US], 1978)	/	/	/	Natural gas	Gaseous	[23]
Study addressing the issue of asphyxiation by natural gas	/	/	/	Natural gas	Gaseous	[13]

3. APPLICATION OF THE APPROACH

As demonstrated by Paltrinieri et al. [5,6], the relative lack of experience in the management of emerging technologies could potentially lead to atypical accident scenarios. An example of this may be represented by LNG regasification technologies, which are playing an increasingly important role in the energy industry [7−9]. For this reason, DyPASI was applied to the most common typologies of LNG regasification terminals. For the sake of brevity, only a representative application is shown in the following: the bow-tie diagram concerning the spherical Moss storage tanks of a floating storage regasification unit (FSRU).

3.1 Bow-Tie Analysis

DyPASI was coupled with the MIMAH [2] methodology to perform a preliminary bow-tie analysis. The latter technique gave an overview of the main hazards related to the type of equipment considered (spherical Moss storage tank). For this category of equipment, several types of loss of containments (LOCs) are identified (breach of shell in vapor and liquid phase and catastrophic rupture). Bow-tie diagrams were built for each LOC and were considered critical events [10].

3.2 DyPASI Application

Risk notions retrieved through the step 1 of DyPASI are reported in Table 4.4 and address the following potential events:

- rapid phase transition, ie, spontaneous and explosive boiling of LNG in contact with water
- boiling liquid expanding vapor explosion (BLEVE) when pressurized LNG undergoes a nearly instantaneous failure
- terrorist attack, because LNG facilities may become targets of piracy or future terrorist attacks

Cryogenic burn, cryogenic damage, and asphyxiation related to the specific properties of LNG.

Table 4.5 shows a summary of the prioritization process results (step 2 of DyPASI application). The main priority is given to terrorist attacks, for which DyPASI collected

Table 4.5 Risk Notions Collected Through the Application of DyPASI [5,22,23]

	Event Typology	No. of DyPASI Records	Severity
1	Terrorist attack	3	Disaster
2	Asphyxiation	2	Accident
3	Boiling liquid expanding vapor explosion	1	Accident
4	Rapid phase transition	8	Incident
5	Cryogenic burns	1	Incident
6	Cryogenic damage	1	Mishap

Table 4.6 Cause–Consequence Chains Describing the Atypical Incident Scenarios Identified [5,11,12]

	Terrorist Attack	Asphyxiation	Rapid Phase Transition	Cryogenic Burns	Cryogenic Damage
Causes	✓ ✓ ✓ ✓	Asphyxiation; High gas concentration; Gas dispersion; Pool formation or gas jet	Overpressure; Rapid phase transition; Rapid heat exchange; Pool formation	Cryogenic burns; Cryogenic liquid release	✓ ✓ ✓ ✓
Critical event	Large breach or catastrophic rupture	Natural gas/liquefied natural gas leak	Liquefied natural gas leak	Liquefied natural gas leak	Breach of shell or catastrophic rupture; Brittle rupture; Brittle structure and impact; Low temperature; Leak of cryogenic liquid (domino effect)
Consequences	Excessive mechanical stress; External impact; Malicious intervention; Terrorist attack	✓ ✓ ✓ ✓	✓ ✓ ✓ ✓		

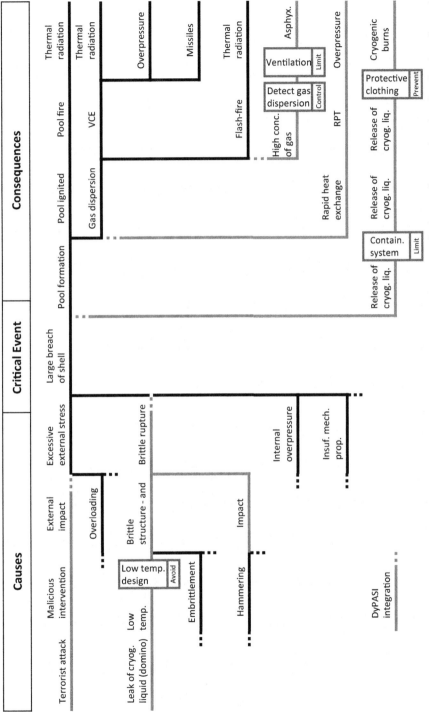

Figure 4.3 Bow-tie diagram concerning a large breach of shell in the liquid phase of an a liquefied natural gas tank with safety barriers. Several fault tree branches have been omitted; they are represented by dotted lines.

Table 4.7 Complete List of the Identified Safety Barriers for the New Diagram Branches Integrated Through DyPASI [5,11,12]

	Hazardous Event	Safety Function	Safety Barrier
Terrorist attack	Terrorist attack	To prevent	Surveillance
	Malicious intervention	To prevent	Security zones
	External impact	To prevent	Control of ship
		To limit	Absorbing barriers
Cryogenic damages	Leak of cryogenic liquid (domino effect)	To avoid	Plant design (distances between equipment)
		To prevent	General leak prevention, control, and limiting measures
		To control	
		To limit	
	Low temperature	To prevent	Inspection
	Brittle structure	To avoid	Low-temperature design
	Impact	To limit	Protect the structure
Cryogenic burns	Release of cryogenic liquid	To prevent	General leak prevention, control, and limiting measures (eg, containment system)
		To control	
		To limit	
	Cryogenic burns	To prevent	Protective clothing
Rapid phase transition	Rapid heat exchange	To prevent	Containment system to prevent water contact
Asphyxiation	High concentration of gas	To control	Detection of gas dispersion
		To limit	Ventilation

three risk notion records and which can potentially lead to multiple fatalities and extensive damage to property.

Table 4.6 shows the cause-consequence chains created on the basis of the early warnings gathered and by means of the "why tree" technique [24] (step 3 of DyPASI application: atypical scenario identification). The only scenario that is not represented is BLEVE, because the MIMAH technique identifies the possibility of BLEVE for LNG as a consequence of a domino effect [25,26]. Domino effects are not explicitly conveyed on the bow-tie diagrams, but escalation resulting from a dangerous phenomenon may be captured in a second bow-tie. For example, a pool fire (a dangerous phenomenon) may be the cause of the catastrophic rupture (BLEVE) of a pressure vessel.

The final result of the integration process is shown in Fig. 4.3 for a spherical Moss storage tank of an FSRU: a bow-tie diagram concerning a large breach of shell in the liquid phase of an LNG tank, which also includes the atypical accident scenarios identified. The atypical scenario elements integrated into the diagram are highlighted. Where possible, events already present in the diagram were exploited to describe atypical scenarios.

The search of risk notions in this case allowed simultaneous identification of specific safety measures for atypical or specific events included within the hazard identification analysis (eg, those proposed by the Sandia report [13]), as required by step 4 of DyPASI application. The safety measures identified have been translated into graphic barriers that were positioned on the diagrams. Safety barriers were classified as actions to avoid, prevent, control, or limit their reference event [26]. A list of safety barriers suggested by the same sources of information as the early warnings collected is shown in Fig. 4.3. This is a confirmation of the importance of risk notion search. A more complete list of the identified safety barriers for the new diagram branches is reported in Table 4.7.

4. CONCLUSIONS

This chapter illustrates a step-by-step tutorial for the application of the DyPASI technique for hazard identification. DyPASI is specifically designed to provide a comprehensive overview of hazards in an industrial process by means of a systematic approach. Its structure allows for analysis reiteration and continuous improvement once new information is available to the analyst.

A representative application of DyPASI was carried out on LNG regasification technologies. This is a challenging case for hazard identification: although related risks can be well known to academics and experts, safety professionals may disregard scenarios for which there is relatively limited experience [5]. DyPASI resulted in providing a significant integration of bow-tie diagrams obtained by the application of conventional static techniques such as the MIMAH approach.

REFERENCES

[1] CCPS — Center for Chemical Process Safety. Guidelines for chemical process quantitative risk analysis. New York, USA: America Institute of Chemical Engineers (AIChE); 2000.
[2] Delvosalle C, Fievez C, Pipart A, Debray B. ARAMIS project: a comprehensive methodology for the identification of reference accident scenarios in process industries. Journal of Hazardous Materials 2006;130:200–19.
[3] Paltrinieri N, Tugnoli A, Buston J, Wardman M, Cozzani V. Dynamic procedure for atypical scenarios identification (DyPASI): a new systematic HAZID tool. Journal of Loss Prevention in the Process Industries 2013;26:683–95.
[4] Zhao J, Cui L, Zhao L, Qiu T, Chen B. Learning HAZOP expert system by case-based reasoning and ontology. Computers & Chemical Engineering 2009;33:371–8.
[5] Paltrinieri N, Tugnoli A, Cozzani V. Hazard identification for innovative LNG regasification technologies. Reliability Engineering and System Safety 2015;137:18–28.

[6]   Paltrinieri N, Wilday J, Wardman M, Cozzani V. Surface installations intended for Carbon Capture and Sequestration: Atypical accident scenarios and their identification. Process Safety and Environmental Protection 2014;92:93–107.

[7]   Energy Information Administration — Office of Integrated Analysis and Forecasting. International energy outlook. (Washington DC): U.S. Department of Energy; 2013.

[8]   Jensen Associates. The outlook for global trade in liquefied natural gas projections to the year 2020. 2007. Weston (MA).

[9]   EUROSTAT. Energy, transport and environment indicators. 2010 ed. Luxemburg: Publications Office of the European Union; 2013.

[10] Uguccioni G. Reference solution containing documents, methods and tools, for the assessment and management of emerging risks related to new and intensified technologies available for LNG regasification terminals. 2010. Stuttgart.

[11] Tugnoli A, Paltrinieri N, Landucci G, Cozzani V. LNG regasification terminals: comparing the inherent safety performance of innovative technologies. Chemical Engineering Transactions 2010: 391–6.

[12] Paltrinieri N, Tugnoli A, Bonvicini S, Cozzani V. Atypical scenarios identification by the DyPASI procedure: application to LNG. Chemical Engineering Transactions 2011:1171–6.

[13] Sandia National Laboratories (SNL). Guidance on risk analysis and safety implications of a large liquefied natural gas (LNG) spill over water. 2004. Albuquerque (NM).

[14] European Commission for Standardization. EN1160, installations and equipment for liquefied natural gas — general characteristics of liquefied natural gas. CEN; 1996.

[15] Bubbico R, Salzano E. Acoustic analysis of blast waves produced by rapid phase transition of LNG released on water. Safety Science 2009;47:515–21.

[16] US EPA. Cabrillo Port Liquefied Natural Gas Deepwater Port Final EIS/EIR — Appendix C3-1. San Francisco: Environmental Protection Agency Region IX; 2007.

[17] CH IV. Safety History of International LNG Operations — Technical Document. Hanover, USA: CH•IV International; 2006.

[18] Planas-Cuchi E, Gasulla N, Ventosa A, Casal J. Explosion of a road tanker containing liquefied natural gas. Journal of Loss Prevention in the Process Industries 2004;17:315–21.

[19] CRS Report for Congress. Maritime Security: Potential Terrorist Attacks and Protection Priorities. (Washington DC): Federation of America Scientists (FAS); 2007.

[20] Husick L, Gale S. Planning a sea-borne terrorist attack. Philadelphia: Foreign Policy Research Institute (FPRI); 2005.

[21] US Government Accountability Office (GAO). Maritime Security: Public Safety Consequences of a Terrorist Attack on a Tanker Carrying Liquefied Natural Gas Need Clarification. (Washington DC): GAO; 2007.

[22] Woodward J, Pitbaldo R. LNG risk based safety: modelling and consequence analysis. 2010. New Jersey, USA.

[23] Technology AEA. Major Hazards Assessment Unit. MHIDAS — Major Hazard Incident Data Service. Bootle: Health and Safety Executive; 2003.

[24] CCPS — Center for Chemical Process Safety. Guidelines for investigating chemical process incidents. New York: America Institute of Chemical Engineers (AIChE); 2003.

[25] Lees FP. Loss prevention in the process industries. UK: Oxford; 1996.

[26] Delvosalle C, Fievez C, Pipart A. ARAMIS (Accidental Risk Assessment Methodology for Industries) — Deliverable D.1.C. 2004. WP1. Mons: Faculté Polytechnique de Mons, Major Risk Research Centre.

Analysis of Initiating Events

CHAPTER 5

Reactive Approaches of Probability Update Based on Bayesian Methods

N. Khakzad[1], H. Yu[2], N. Paltrinieri[3,4], F. Khan[5]

[1]Delft University of Technology, Delft, The Netherlands; [2]University of Tasmania, Launceston, Australia; [3]Norwegian University of Science and Technology (NTNU), Trondheim, Norway; [4]SINTEF Technology and Society, Trondheim, Norway; [5]Memorial University of Newfoundland, St John's, NL, Canada

1. INTRODUCTION

Safety analysis in hazardous installations such as chemical and process plants is crucial owing to the presence of large inventories of hazardous materials, which are usually stored and handled in high-pressure, high-temperature conditions. This, along with the large number of process equipment, such as compressors, separators, heat exchangers, and complex pipeline networks, can turn any undesired release of contained hazardous substances into catastrophic fires, explosions, or dispersions of toxic gases. There have been many catastrophic accidents within the chemical industry over the past decades. These could have been prevented if near-accident data (accident precursors, such as near misses and mishaps) were incorporated and analyzed in a dynamic safety analysis framework to forecast likely accident scenarios. For example, before the blowout of Macondo well on April 20, 2010, in the Gulf of Mexico, several kicks and lost circulations were experienced during the drilling phase in February, March, and April of the same year [1].

Thus, it is important to broaden the scope of safety risk analysis by considering not only the accident precursors but also varying process parameters (temperature, pressure, flow, etc.) so that the probability of malfunctions and accidents can be predicted and continuously updated in a real-time fashion. Many techniques have been developed for accident scenario modeling and safety assessment, among which fault tree analysis (FTA), event tree analysis (ETA), and bow-tie analysis (BTA) are very popular. Standard FTA, ETA, and BTA are not suitable for analyzing large systems, particularly if the system includes redundant components or common-cause failures or presents dynamic behavior or time-varying parameters [2].

In recent years, Bayesian analysis and, particularly, the Bayesian network (BN) method have widely been used in safety assessment and management of chemical establishments. Generally speaking, in Bayesian methods the accident precursor data are used in the form of a likelihood function via Bayes' theorem to update the analyst's prior belief about accident probability or the failure probability of safety barriers [3—6]. A safety

Dynamic Risk Analysis in the Chemical and Petroleum Industry
ISBN 978-0-12-803765-2

barrier is a passive or active, physical, technical, operational, or human system or function aimed at preventing, controlling, or mitigating undesired events or accidents [7]. BN is a graphical probabilistic technique based on Bayes' theorem for reasoning under uncertainty and data scarcity. BNs can be used either in a forward analysis to predict the probability of unknown variables (eg, likely failure of a safety barrier) based on known variables (eg, degradation or failure of the safety barrier components) or in a backward analysis to update the probability of known variables given some evidence (eg, to update the failure probability of a safety component given the failure of the safety system) based on Bayes' theorem. Owing to its flexible graphical representation and robust reasoning engine, BN has proved a promising methodology for safety assessment of a wide variety of process equipment and plants [2,8−11]. A comprehensive statistical review of BN application can be found in Weber et al. [12], in which the growing appeal of BN in various areas of reliability engineering, safety risk analysis, and maintenance engineering has been shown over the past decade.

2. BAYESIAN INFERENCE

2.1 Bayes' Theorem

A major task in any dynamic safety risk analysis is first to predict *a priori* probability distribution for a variable of interest θ (eg, a safety barrier probability of failure) and then update the prior belief given the observation(s) of other variables, also known as evidence E (eg, the failure of other safety barriers or components). The prior probability distribution can be estimated in several ways including statistical analysis of historical data or data collected from inspection/condition monitoring, deductive reasoning by means of quantitative risk analysis techniques (eg, FTA, ETA, and BTA), or expert judgment. The updating of the prior probability distribution, however, is usually performed based on the application of Bayes' theorem

$$f_1(\theta|E) = \frac{f_0(\theta)L(E|\theta)}{\int f_0(\theta)L(E|\theta)d\theta} \tag{5.1}$$

where $f_1(\theta|E)$ is the updated (posterior) probability distribution of θ given the evidence E; $f_0(\theta)$ is the prior probability distribution of θ; and $L(E|\theta)$ is the likelihood function of θ or alternatively the probability distribution of the evidence given the parameter θ. In the case of there being discrete probability values instead of probability distributions, Eq. (5.1) can be simplified as:

$$P(\theta|E) = \frac{P(\theta)L(E|\theta)}{\sum_\theta P(\theta)L(E|\theta)} \tag{5.2}$$

Because in Eq. (5.1) (and also Eq. (5.2)) the denominator is used to normalize the posterior probability distribution, in many applications it suffices to employ Eq. (5.1) in the form of $f_1(\theta|E) \propto f_0(\theta)L(E|\theta)$, implying that the posterior probability distribution is proportionate to the product of the prior and the likelihood probability distributions. Provided that the prior probability distribution and the likelihood function are of the conjugate distributions, eg, beta-binomial or gamma-Poisson distributions, the posterior probability will adopt the same probability distribution of the prior probability yet with different (updated) distribution parameters [3,4]. For example, consider a case where the prior probability of failure of a safety barrier follows a beta distribution $f_0(\theta) \propto \theta^\alpha(1-\theta)^\beta$ and the likelihood function can be estimated using a binomial distribution $L(E|\theta) \propto \theta^m(1-\theta)^{n-m}$. The parameters of beta distribution are α and β, whereas m and n are the parameters of binomial distribution, indicating that the safety barrier has failed m of n times when demanded. As a result, the posterior probability distribution of θ will be $f_1(\theta|E) \propto \theta^{\alpha+m}(1-\theta)^{\beta+n-m}$, which is a beta distribution with the updated parameters $\alpha' = \alpha + m$ and $\beta' = \beta + n - m$. In the case of nonconjugate distributions, however, numerical methods such as Markov chain Monte Carlo sampling can be employed to derive the posterior probability distribution [5,6].

2.2 Hierarchical Bayesian Analysis

A Bayesian analysis can be extended as a hierarchical Bayesian analysis (HBA) to better capture the uncertainty of prior probability distributions. In HBA, a generic prior distribution is developed for the parameter of interest θ based on available (usually scarce) data or expert opinion. This generic prior distribution can be updated to yield a more case-specific informative prior distribution given any evidence in the form of likelihood function. In this way, for example, the generic failure probabilities of safety barriers retrieved from the literature or relevant databases can be adjusted to specific applications on the basis of the actual safety barrier performance.

To develop the informative prior distribution, θ is considered to follow a generic distribution with its own parameters α and β as $f_g(\theta|\alpha,\beta)$; α and β are usually referred to as hyperparameters. To cope with the uncertainty arising from either the data scarcity or the analyst's lack of knowledge about the prior distribution of θ, usually diffusive or noninformative prior distributions (eg, various forms of uniform distribution) are used for α and β. These noninformative prior distributions of α and β are also known as hyperprior distributions $\pi_0(\alpha,\beta)$. With a set of observations, E, the hyperprior distributions can be updated to yield hyperposterior distributions $\pi_1(\alpha,\beta|E)$:

$$\pi_1(\alpha, \beta|E) = \frac{\pi_0(\alpha, \beta)L(E|\alpha, \beta)}{\iint \pi_0(\alpha, \beta)L(E|\alpha, \beta)\, d\alpha\, d\beta} \tag{5.3}$$

With the hyperposterior distributions of α and β obtained by Eq. (5.3), the informative prior distribution of θ can be estimated as

$$f_0(\theta|E) = \iint f_g(\theta|\alpha, \beta)\pi_1(\alpha, \beta|E) \, d\alpha \, d\beta \tag{5.4}$$

$f_0(\theta|E)$ is the informative prior distribution of θ, which can be further updated to render a posterior distribution given more observations E' via the employment of Bayes' theorem:

$$f_1(\theta|E') = \frac{f_0(\theta|E)L(E'|\theta)}{\int f_0(\theta|E)L(E'|\theta)d\theta} \tag{5.5}$$

3. BAYESIAN NETWORK

3.1 Conventional Bayesian Network

A BN is a directed acyclic graph for reasoning under uncertainty [13] wherein the nodes represent random variables (eg, process parameters or function/malfunction of safety barriers) and are connected by means of directed arcs. The arcs denote dependencies and in most cases causal relationships between the linked nodes. Accordingly, the conditional probability tables (CPTs) assigned to the nodes determine the type and strength of such dependencies (similar to AND/OR gate in fault trees). In a BN, nodes from which arcs are directed are called "parent nodes," whereas nodes to which arcs are directed are called "child nodes." As a result, a node can simultaneously be the child of a node and the parent of another node. The nodes with no parent are called "root nodes," and the nodes with no children are called "leaf nodes"; the rest of nodes are "intermediate nodes."

Fig. 5.1 depicts a typical BN comprising four nodes, where $X1$ is the root node; $X2$ is the intermediate node; and $X3$ and $X4$ are leaf nodes. Furthermore, $X2$ is both the child of $X1$ and the parent of $X3$ and $X4$. For the sake of exemplification, assume that the BN in Fig. 5.1 can be used for failure analysis of a water sprinkler system, comprising a smoke (flame) detector ($X1$), an actuator ($X2$), an alarm ($X3$), and a water sprinkler ($X4$). Given a fire, the smoke detector can activate the actuator ($X1 \rightarrow X2$) and ring the alarm ($X1 \rightarrow X3$); the actuator then would trigger the water sprinkler ($X2 \rightarrow X4$). To increase the reliability of the system, the actuator can also trigger the alarm ($X2 \rightarrow X3$). Using the chain rule and the D-separation criterion [13], BN factorizes the joint probability distribution of a set of random variables (nodes) $U = \{X1, X2, ..., Xn\}$ as the multiplication of the probabilities of the nodes given their immediate parents just by considering local dependencies

$$P(U) = \prod_{i=1}^{n} P(Xi|\mathrm{Pa}(Xi)) \tag{5.6}$$

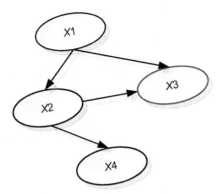

Figure 5.1 A typical Bayesian network. *X1* is the root node; *X2* is the intermediate node; and *X3* and *X4* are leaf nodes.

where $P(U)$ is the joint probability distribution of variables and $\mathrm{Pa}(Xi)$ is the parent set of variable Xi.

For instance, the joint probability distribution of the random variables $X1$, $X2$, $X3$, and $X4$ in the BN of Fig. 5.1 can exclusively be expanded as $P(X1,X2,X3,X4) = P(X1)P(X2|X1)P(X3|X1,X2)P(X4|X2)$. Accordingly, the marginal probability of each random variable, for example, the failure probability $X3$, can be calculated via marginalization as: $P(X3) = \sum_{X1,X2,X4}P(X1, X2, X3, X4) = \sum_{X1}\left(P(X1)\sum_{X2}\left(P(X2|X1)P(X3|X1, X2)\sum_{X4}P(X4|X2)\right)\right)$.

The main application of a BN in dynamic safety assessment is to update prior probabilities (eg, the failure probabilities of the components of a safety system) given a new set of observations or evidence (eg, the failure of the system or some components). A BN directly takes advantage of Bayes' theorem in Eq. (5.2) to update the probabilities:

$$P(U|E) = \frac{P(U)P(E|U)}{\sum_U P(U)P(E|U)} = \frac{P(U, E)}{\sum_U P(U, E)} \tag{5.7}$$

For example, assuming binary random variables (eg, $X4 = \{x4 : \text{function}, \overline{x4} : \text{malfunction}\}$) and knowing that the random variable $X4$ is in the state $\overline{x4}$: malfunction, the updated probability of $X1$ being in the state $x1$: function can be calculated as $P(X1 = x1|X4 = \overline{x4}) = \frac{\sum_{X2,X3}P(x1,X2,X3,\overline{x4})}{\sum_{X1,X2,X3}P(X1,X2,X3,\overline{x4})}$.

In safety risk analysis of chemical and process facilities, the probability updating feature of a BN can effectively be employed to identify the root causes or failure modes (eg, the components of the safety system in Fig. 5.1) contributing the most to the failure of a system of interest (eg, the sprinkler system in Fig. 5.1) [2,8–10]. In this regard, not only the events contributing the most but also the *most probable configuration* of events leading to the failure of the safety system can be determined [2].

3.2 Dynamic Bayesian Network

A dynamic Bayesian network (DBNtrun-1) [13] offers a flexible structure and unique modeling techniques to intuitively model temporal dependencies via a probabilistic framework. The reasoning and learning engine of a DBN allows time dependencies and complex interactions among the components of a safety system to be taken into account. This attribute of a DBN helps a DBN outperform a conventional BN in likelihood modeling and safety analysis of stochastic systems. Generally speaking, two different approaches have been developed for a DBN: (1) interval-based approach and (2) instant-based approach.

3.2.1 Interval-based Dynamic Bayesian Network

In this approach, the timeline $[0, +\infty)$ is partitioned into $n + 1$ time intervals, each as a state of the random variable of interest. That is, the mission time $[0, T]$ is divided into n intervals (n states), whereas the last time interval $(T, +\infty)$ is left as the state $n + 1$ [14]. As a result, each random variable (node) in the DBN has a finite number of states equal to the number of time intervals. For example, if random variable $X1$ in Fig. 5.1 is mentioned to be in the ith state ($1 \leq i \leq n$) or simply $X1 = i$, it means $X1$ has failed in the ith interval (Fig. 5.2). In other words, time to failure of $X1$ can be shown as $t_{X1} \in ((i-1)\Delta, i\Delta]$. Thus

$$P(X1 = i) = P((i-1)\Delta < t_{X1} < i\Delta) = \int_{(i-1)\Delta}^{i\Delta} f_{X1}(t) \, dt$$

$$= F_{X1}(i\Delta) - F_{X1}((i-1)\Delta) \tag{5.8}$$

where t_{X1} is the time to failure of $X1$; $f_{X1}(\cdot)$ is the probability distribution function of t_{X1}; $F_{X1}(\cdot)$ is the cumulative distribution function of t_{X1}; Δ is the interval length $\Delta = \frac{T}{n}$, and n is the time granularity [14]. Similarly, if $X1$ is said to be in the $(n + 1)^{th}$ state, this means $X1$ has not failed during the mission time T:

$$P(X1 = n+1) = P(t_{X1} > T) = \int_{T}^{\infty} f_{X1}(t)dt = 1 - F_{X1}(T) \tag{5.9}$$

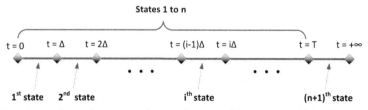

Figure 5.2 Time intervals as the states of a random variable in interval-based dynamic Bayesian network [15].

As mentioned previously, the application of interval-based DBN facilitates the incorporation of temporal dependencies in the failure analysis of safety systems. As a result, the probability of a sequential failure of safety barriers (eg, the safety barriers in an ETA or Swiss Cheese model) can be modeled rather than merely the failure probability of a set of safety barriers, ignoring the time order of failures. An interval-based DBN is relatively simple to construct and can be solved using standard inference algorithms. However, it has been criticized for resulting in large CPTs if the timeline is to be divided into finer time intervals to achieve a better degree of accuracy. More detailed explanation and applications of this approach to safety assessment of process systems can be found in the literature [14–16].

3.2.2 Instant-based Dynamic Bayesian Network

Similar to the previous approach, in the instant-based DBN, the timeline is divided into a finite number of time instants (time slices). However, instead of the calculation of the failure probabilities of safety barriers in each time interval, identical BN structures (such as the one depicted in Fig. 5.1) are generated at each time instant and connected to each other by means of temporal arcs (Fig. 5.3).

Using the instant-based DBN, a node at i^{th} time slice can be conditionally dependent not only on its parents at the same time slice but also on its parents and even itself at previous time slices. However, if the temporal transition rates (or transition probabilities) are

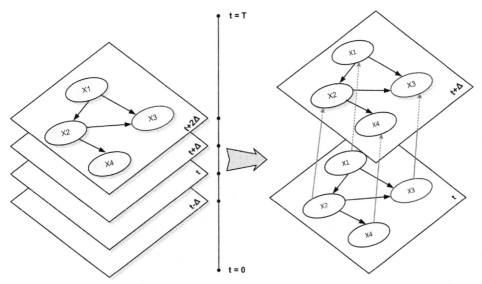

Figure 5.3 Schematic of an instant-based dynamic Bayesian network by replicating the same ordinary Bayesian networks at consecutive time slices. The *arcs in red* (gray in print versions) represent temporal arcs [17].

constant, only two time slices can be considered in the modeling [18]. As such, the joint probability distribution of a set of random variables (eg, the failure probability of a set of safety barriers) at time $t + \Delta$ can be expanded as [17]

$$P\left(U^{t+\Delta}\right) = P\left(X1^{t+\Delta}, X2^{t+\Delta}, ..., Xn^{t+\Delta}\right) = \prod_{i=1}^{n} P\left(Xi^{t+\Delta} \big| Xi^{t}, \mathrm{Pa}(Xi^{t}), \mathrm{Pa}\left(Xi^{t+\Delta}\right)\right)$$

(5.10)

where Xi^{t} and $Xi^{t+\Delta}$ are the copies of Xi in two consecutive time slices with a time interval of Δ, and $\mathrm{Pa}(Xi^{t})$ and $\mathrm{Pa}(Xi^{t+\Delta})$ are the parent sets of Xi at the foregoing time slices. The directed arcs connecting the nodes in the same time slice are called "intraslice arcs" (black arcs in Fig. 5.3), whereas the arcs linking the nodes in consecutive time slices are called "temporal arcs" or "interslice arcs" [red arcs (gray in print versions) in Fig. 5.3] [18]. For example, in Fig. 5.3, the conditional probability of the copy of $X4$ at $t + \Delta$ can be calculated as $P(X4^{t+\Delta}|X4^{t}, X2^{t+\Delta})$. Instant-based DBN can effectively be employed in the failure analysis of safety barriers that degrade over time (eg, degradation of fireproof coating exposed to heat radiation) or to capture the influence of maintenance on the failure probability of barriers. In the case of the water sprinkler system displayed in Fig. 5.1, for example, the depletion of the available water resource as a function of time can be modeled as $(X4^{t} \rightarrow X4^{t+\Delta t})$ in the DBN of Fig. 5.3. Some applications of instant-based DBN can be found in the literature [17–18].

4. LIMITED MEMORY INFLUENCE DIAGRAM

A BN can be extended to a limited memory influence diagram (LIMID) using two additional types of nodes, decision and utility nodes (Fig. 5.4). Each decision node has a finite set of decision alternatives as its states. A decision node should be the parent of all those chance nodes for which probability distributions depend on at least one of the decision alternatives (eg, node $X3$ in Fig. 5.4). Likewise, the decision node should be the child of all those chance nodes with states that have to be known to the decision-maker prior to making that decision (eg, node $X1$ in Fig. 5.4). For example, considering the water sprinkler system in Fig. 5.1, the decision node in Fig. 5.4 can consist of the decision alternatives to (1) activate the alarm ($X3$) given that the smoke detector ($X1$) has detected a fire, or (2) not activate the alarm ($X3$) even though the smoke detector ($X1$) has detected a fire (because of a safe-to-fail[1] state of $X1$). Decision nodes are conventionally presented as rectangles, whereas utility nodes are shown as diamonds.

---

[1] In a safe-to-fail state, the smoke detector erroneously indicates a fire which does not exist.

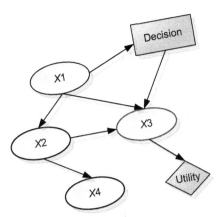

Figure 5.4 A limited memory influence diagram formed by adding decision node (rectangle) and utility node (diamond) to the Bayesian network of Fig. 5.1.

A utility node is a random variable with values (utility values) that express the preferences of the decision-maker regarding the outcomes of the decision to be made [eg, the cost attributed to the unnecessary evacuation of the building when the alarm goes off because of an error in the smoke detector or the cost (risk) attributed to the late evacuation of the building]. Each utility node is assigned a utility table of values that are not probabilities (unlike CPT) but are rather numeric values (positive or negative) determined by the decision-maker for each configuration of parent nodes [19]. In Fig. 5.4, considering a set of n mutually exclusive decision alternatives for the decision node as Decision $= \{d_1, d_2, ..., d_n\}$ and m states for node $X3$ as $X3 = \{x3_1, x3_2, ..., x3_m\}$, the utility table for the utility node includes $n \times m$ utility values $u_{ij} = U(d_i, x3_j)$ for combinations of the decision alternatives and the states of $X3$. Accordingly, the expected utility of the ith decision alternative, $EU(d_i)$, can be calculated as:

$$EU(d_i) = \sum_{X3} P(X3|d_i) U(d_i, X3)$$
$$= P(x3_1|d_i)u_{i1} + P(x3_2|d_i)u_{i2} + \cdots + P(x3_m|d_i)u_{im} \qquad (5.11)$$

As a result, the decision alternative with the maximum expected utility can be selected as the optimal decision. Utility values are usually determined by consulting experts, considering the preferences of the decision-maker. Utility values can also be generated using appropriate utility functions [19]. A recent application of LIMID to safety analysis of chemical plants can be found in Khakzad [20].

5. CONCLUSIONS

In this chapter we introduced state-of-the-art applications of Bayesian analysis and BNs to predict and update the failure probability of safety barriers in chemical facilities in a reactive framework. As discussed, the reason for the popularity of BNs in dynamic safety analysis lies in the fact that it benefits from both qualitative and quantitative modeling techniques. From a qualitative perspective, a BN takes advantage of a highly flexible structure to incorporate the causal relationships and dependencies among the components of a safety system. From a quantitative perspective, a BN is able to explicitly determine the type and strength of such dependencies by means of conditional probabilities. The unique probability updating feature of a BN is especially of great importance when the near-accident data and equipment malfunction observations can be employed to update the failure probabilities in a real-time fashion. We also demonstrated that the versatile structure and robust reasoning engine of a BN can effectively be taken into account in risk-based decision analysis by means of decision and utility nodes.

REFERENCES

[1] Skogdalen JE, Vinnem JE. Combining precursor incidents investigations and QRA in oil and gas industry. Reliability Engineering and System Safety 2012;101:48—58.
[2] Khakzad N, Khan F, Amyotte P. Safety analysis in process facilities: comparison of fault tree and Bayesian network approaches. Reliability Engineering and System Safety 2011;96:925—32.
[3] Meel A, Seider WD. Plant-specific dynamic failure assessment using Bayesian theory. Chemical Engineering Science 2006;61(21):7036—56.
[4] Khakzad N, Khan F, Amyotte P. Dynamic risk analysis using bow-tie approach. Reliability Engineering and System Safety 2012;104:36—44.
[5] Khakzad N, Khan F, Paltrinieri N. On the application of near accident data to risk analysis of major accidents. Reliability Engineering and System Safety 2014;126:116—25.
[6] Khakzad N, Khakzad S, Khan F. Probabilistic risk assessment of major accidents: application to offshore blowouts in the Gulf of Mexico. Natural Hazards 2014;74:1759—71.
[7] Sklet S. Safety barriers: definition, classification, and performance. Journal of Loss Prevention in the Process Industries 2006;19:494—506.
[8] Khakzad N, Khan F, Amyotte P. Dynamic safety analysis of process systems by mapping bow-tie into Bayesian network. Process Safety and Environmental Protection 2013;91:46—53.
[9] Khakzad N, Khan F, Amyotte P. Quantitative risk analysis of offshore drilling operations: a Bayesian approach. Safety Science 2013;57:108—17.
[10] Khakzad N, Khan F, Amyotte P, Cozzani V. Domino effect analysis using Bayesian networks. Risk Analysis 2013;33(2):292—306.
[11] Khakzad N, Reniers G. Risk-based design of process plants with regard to domino effects and land use planning. Journal of Hazardous Materials 2015;299:289—97.
[12] Weber P, Medina-Oliva G, Simon C, Iung B. Overview on Bayesian networks applications for dependability, risk analysis and maintenance areas. Engineering Application of Artificial Intelligence 2010;25:671—82.
[13] Neapolitan R. Learning Bayesian networks. 1st ed. Upper Saddle River, NJ: Prentice Hall, Inc.; 2003.
[14] Boudali H, Dugan JB. A discrete-time Bayesian network reliability modeling and analysis framework. Reliability Engineering and System Safety 2005;87:337—49.
[15] Khakzad N, Khan F, Amyotte P. Risk-based design of process systems using discrete-time Bayesian networks. Reliability Engineering and System Safety 2013;109:5—17.
[16] Khakzad N, Khan F, Amyotte P, Cozzani V. Risk management of domino effects considering dynamic consequence analysis. Risk Analysis 2014;34(6):1128—38.
[17] Khakzad N. Application of dynamic Bayesian network to risk analysis of domino effects in chemical infrastructures. Reliability Engineering and System Safety 2015;138:263—72.

[18] Montani S, Portinale L, Bobbio A, Codetta-Raiteri D. RADYBAN: a tool for reliability analysis of dynamic fault trees through conversion into dynamic Bayesian networks. Reliability Engineering and System Safety 2008;93:922—32.
[19] Gilboa I. Theory of decision under uncertainty. New York, NY: Cambridge University Press; 2009.
[20] Khakzad N, Reniers G. Cost-effective allocation of safety measures in chemical plants w.r.t land-use planning. Safety Science 2015. http://dx.doi.org/10.1016/j.ssci.2015.10.010.

CHAPTER 6

Proactive Approaches of Dynamic Risk Assessment Based on Indicators

N. Paltrinieri[1,2], G. Landucci[3], W.R. Nelson[4], S. Hauge[2]
[1]Norwegian University of Science and Technology (NTNU), Trondheim, Norway; [2]SINTEF Technology and Society, Trondheim, Norway; [3]University of Pisa, Pisa, Italy; [4]DNV GL, Houston, TX, United States

1. INTRODUCTION

Prevention of major accidents in the chemical and process industries requires both continuous learning from events occurring around us and tireless surveillance of critical safety systems. This can be achieved by using appropriate indicators and effectively processing the information they convey. In fact, in several major accidents, hindsight has shown that if the available early warnings had been detected and managed in advance, the unwanted events could have been avoided, as shown by Paltrinieri et al. for the accident that occurred at Buncefield in 2005 [1]. For this reason, this chapter addresses approaches of dynamic risk assessment based on proactive indicators, which aim to enable recognition of such signals in time to reduce major accident risk. Techniques for the development of safety or risk indicators are first presented, followed by descriptions of methodologies for aggregating and relating such indicators to the overall risk level. The terms "safety" and "risk" clearly refer to different concepts (protection from and exposure to potential accidents, respectively), and indicators may equally refer to either side of the problem, unless they are related to the overall risk, which is usually expressed as the expected number of fatalities within a specific population per year. (Other risk metrics are also presented elsewhere [2].)

2. PROACTIVE AND DYNAMIC FEATURES

The use of safety or risk indicators may allow risk assessment to assume both dynamic and proactive features. Appropriate sets of indicators collected and evaluated on a regular basis can provide information on overall risk level variation. Such continuous assessment is nowadays improved by advanced information technology allowing for real-time data sharing, processing, and visualization of related information and support to decision-making. However, the choice of indicators may affect the ability to control risk (proactivity), rather than just reporting its increase after an unwanted event has occurred (reactivity). The former may be addressed by a set of indicators that are mainly leading, whereas the latter may be addressed by a set of indicators that are mainly lagging.

Dynamic Risk Analysis in the Chemical and Petroleum Industry
ISBN 978-0-12-803765 2

Leading indicators are a form of active monitoring of key events or activities that are essential to deliver the desired safety outcome [3]. They can range from the "maximum number of simultaneous operations last month" to the "number of hours of backlog in preventive maintenance on safety critical equipment." They represent early deviations from the ideal situation that can lead to further escalation of negative consequences. Human and organizational factors often (but not always) represent such underlying causes. For this reason, the adoption of indicators addressing human and organizational factors may enable risk assessment with proactive capabilities. It is often experienced that indicators on technical equipment can be automatically retrieved from online systems, such as maintenance management systems and condition monitoring systems. Indicators on human and organizational factors are generally more difficult to obtain and rely on manual input and assessment.

Lagging indicators are a form of reactive monitoring requiring reporting and investigation of specific incidents and events to discover weaknesses in the system. Lagging indicators show when a desired safety outcome has failed or has not been achieved [3]. They may range from "number of failures of safety critical equipment during last month" or "number of corrosion events detected during last inspection" to "number of accumulated hours of hot work during last week." The use of lagging indicators is essential because they provide important feedback from the system. For this reason they should be coupled with the leading indicators to more comprehensively assess risk.

The approaches to the development of safety/risk indicators covering technical, human, and organizational (THO) causes may be grouped into two classes (classes I and II in Fig. 6.1), as shown in Section 3 of this chapter. Class I is characterized by a retrospective perspective where indicators are developed on the basis of the effect of THO factors in past accidents, and correlation with the overall safety is assumed. However, major accidents are rare events, and the correlation between critical indicators and safety may not be conclusively demonstrated. Class II is characterized by a predictive perspective, where indicators are defined on the basis of risk models [such as quantitative risk analysis (QRA)] for the potential accident scenarios addressed, and the connection to the overall risk level is logically supported by these models.

To accurately assess variations of the overall risk level, which may be expressed with different risk metrics [2], specific techniques aggregating the information provided by the indicators have been defined. Classes III and IV in Fig. 6.1 group such approaches (represented by the methods described in Sections 4 and 5, respectively) and are graphically depicted closer to the risk variation box because they allow more reliable evaluation of risk on a real-time basis. The main difference between the two classes in terms of risk evaluation is in the development and use of indicators. In class III, limited sets of risk indicators may not allow comprehensive coverage of THO factors, whereas the class IV approach employs ad hoc indicators for proactive risk assessment.

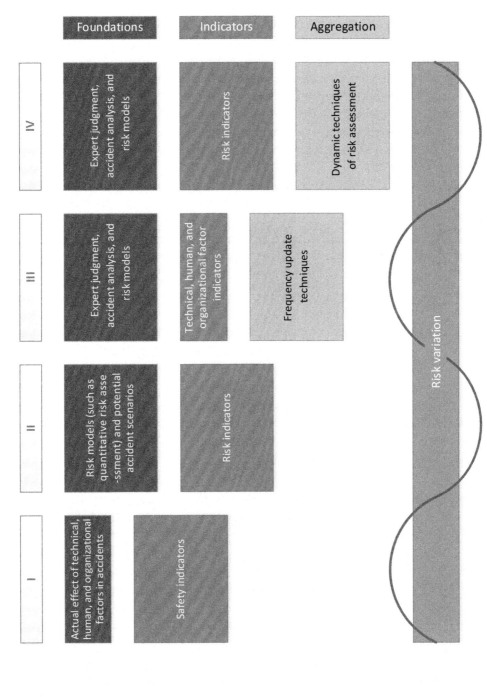

Figure 6.1 Classes of approaches for dynamic risk assessment through monitoring of technical, human, and organizational factors.

3. TECHNIQUES FOR DEVELOPMENT OF INDICATORS

As stated by Øien et al. [11], the research on indicators started with the need to measure safety or risk. The main function of such a measure is to describe the safety/risk level within an organization, establishment, or work unit. In this chapter, safety/risk indicators are defined as follows:

- They should provide numerical values (number or ratio).
- They should be updated at regular intervals.
- They should cover selected determinants of overall safety or risk.

Table 6.1 shows representative approaches for the development of indicators describing THO factors affecting safety or risk. Several of these approaches for the

Table 6.1 Representative Approaches for Development of Technical, Human, and Organizational Indicators

Indicators or Approaches for Their Development	Indicator Typology	Comments
United States Nuclear Regulatory Commission work	Mostly development of indicators measuring the effect of organizational factors on safety (class I in Fig. 6.1), with exceptions representing transition to predictive perspective (class II in Fig. 6.1).	Representative exceptions: Osborn et al. [4,5] define indicators for organizational patterns predicting success and failure, and Boccio et al. [6] and Wreathall et al. [7] produce a first attempt of probabilistic-oriented risk-based indicators.
Performance indicators defined by the World Association of Nuclear Operators	Set of performance indicators in the areas of nuclear power plant safety, reliability, efficiency, and personnel safety [8]	For most of the indicators defined, the degree of correlation with safety/risk is unknown (class I) [9].
Operational safety indicators	Set of indicators for surveillance of the operational safety within a power plant covering technical, operational, and organizational conditions [10]	Framework for identifying performance indicators for circumstances related to safety and economic issues is included. The relationship between the indicators and the risk level is unclear (class I) [10].
Safety performance indicators	Safety indicators (class I) developed, but risk analysis is used as framework for identification and	Safety is evaluated by assessing performance level and trend, ie, comparing indicators to reference values and revealing

Table 6.1 Representative Approaches for Development of Technical, Human, and Organizational Indicators—cont'd

Indicators or Approaches for Their Development	Indicator Typology	Comments
	structuring of safety performance areas [8]	indicator increase or decrease [8].
Operator-specific safety indicators	Indicators for both nonphysical systems (organizational and administrative systems) and physical systems (safety systems, ie, equipment) [8]	Positive correlation between indicators and safety is assumed, even though it has not been evaluated (class I). The safety indicators have a potential to improve the experiential learning [8].
Probabilistic indicators	Indicators identified by evaluating trends of events originating from safety functions, systems, and components; relevance assessed through risk analysis [11]	Indicators are developed based on detailed study of equipment events contained in a specific database. Effect on risk is evaluated through specific plant analysis. Connection with safety/risk is known (class II) [11].
Probabilistic safety assessment–based risk indicators	Indicators based on probabilistic safety assessment established at different levels, from level 1 indicators dealing with total facility risk to other sublevels [10]	These kinds of indicators are developed also using quantitative risk analysis as a basis to monitor possible changes in the risk level of oil and gas plants (class II) [12].
Accident sequence precursor–based indicators	Indicators based on accident sequence precursors involving an initiating event, system failure/unavailability, or the combination of the two [13]	Both qualitative investigation of near misses (class I) and quantitative analysis of the events based on risk analysis (class II) are carried out for the definition of indicators [13].
Resilience-based early warning indicators	Indicators based on the capability of coping with the unexpected (resilience), measuring organizational factors as potential for resilience [12]	Such indicators have been developed both in a retrospective way on the basis of past accidents (class I) [1] and in a predictive way on the basis of the risk analysis model used (class II) [14].

development of major hazard indicators were primarily defined for the nuclear power industry. However, the chemical process and petroleum industries have contributed with the definition of specific techniques [11].

4. TECHNIQUES FOR FREQUENCY MODIFICATION

Quantitative risk assessment is usually aimed at representing the initial risk status of a chemical process facility under analysis, in terms of both frequency (F) and expected consequence severity (M, namely "magnitude"). According to the Center for Chemical Process Safety [15], the risk may be expressed as follows:

$$R = F \times M \tag{6.1}$$

Although M is associated with a potential for damage intrinsic in the facility under consideration (considering constant release severity and plant population), F may be subject to remarkable variation due to aging, corrosion, fatigue, poor safety culture, and other dynamic factors. In fact, the frequency term F takes into account the failure of safety systems limiting the occurrence of the accidents. Hence, we may consider risk a time-evolving parameter as follows:

$$R(t) = F(t) \times M_0 \tag{6.2}$$

where M_0 is a constant level of damage and $F(t)$ represents the possible frequency variation in time (t). At the initial time of operation ($t = 0$), when a first QRA is usually performed, we obtain the initial risk level (R_0) as follows:

$$R_0 = F(t = 0) \times M_0 = F_0 \times M_0 \tag{6.3}$$

Combining Eqs. (6.2) and (6.3), we obtain:

$$R(t)/R_0 \cong F(t)/F_0 \equiv \Omega(t) \tag{6.4}$$

Hence, the relative variation of risk level is associated with the modification of the initial frequency value F_0 through a frequency modification factor $\Omega(t)$ that changes during the lifetime of the plant.

The initial frequency value F_0 is usually indicated as "baseline frequency," and it is derived from standard literature databases [16] or obtained through "parts count" [17].

The modification factor $\Omega(t)$ is a parameter updated in time and aimed at considering THO aspects that are related to increase or decrease of the failure likelihood. Several examples of frequency modification factors are available in the literature [15,17–19]. However, none of them have been applied in a time-evolving perspective.

Frequency modification factors are aimed at introducing modification to the baseline frequency value. They usually take into account indicators defined under different forms (often not explicitly referred to as "indicators") describing health, safety, and environmental management; maintenance policy; equipment working conditions; external

Table 6.2 Overview of Literature Methods for Frequency Modification

	Technical Factors	Human and Organizational Factors	References
Center for Chemical Process Safety	Based on expert judgment evaluation	Based on expert judgment evaluation	[15]
American Petroleum Institute 581	Based on equipment and surrounding environment	Based on operator feedback and monitoring of plant conditions	[19]
MANagement Assessment Guidelines in the Evaluation of Risk)	N/A	Based on several managerial aspects	[20]
Barrier	Based on barrier scoring	N/A	[17]

severe environmental impacts; and other relevant issues affecting the baseline frequency value [15,17−19].

Several methods are available in the technical and scientific literature (Table 6.2). These methods can be classified as class III approaches (Fig. 6.1) owing to the relatively limited focus on comprehensive and proactive indicators. In the following, a review of some representative methods is summarized.

The Center for Chemical Process Safety method is a simplified method for variation of baseline frequency data based on expert judgment. An example of application to transport pipelines is provided [15]. This method may lead to significant uncertainties if applied to operating process facilities, where local practices can be quite different [17].

The American Petroleum Institute (API) 581 method is primarily intended to support risk-based inspection programs in refineries, eg, prioritizing the inspections on the basis of the risk associated with each piece of equipment [19]. According to this method, baseline frequency data (F_0) gathered from onshore refining and chemical processing equipment databases are modified as follows:

$$F = F_0 \times EF_1 \times MF_1 \tag{6.5}$$

where F is the modified frequency value, EF_1 is the equipment modification factor, and MF_1 is the management system modification factor. The EF_1 factor is based on technical aspects affecting the likelihood of failure, which are related to equipment features and the environment in which the equipment operates. The MF_1 factor reflects the influence of the facility management system on the mechanical integrity of the plant. MF_1 evaluation is based on interviews with personnel and monitoring of plant conditions. It is worth mentioning that the management evaluation questionnaire can be defined as static because it has not been updated since 2000. What was deemed a good performance then may only be viewed as average today. This leads to a nonconservative estimation of the associated frequencies [17].

The management assessment guidelines in the evaluation of risk method is aimed at determining a site-specific failure frequency based on the generic average failure frequency corrected by a management factor (MF_2) derived from a site-specific assessment procedure [20]. The procedure is based on a questionnaire covering several organizational aspects (eg, procedures, safety culture, incident investigation, and other organizational factors). To each aspect, a qualitative indicator is assigned, ie, average, good, or bad. In case all the indicators are average, good, or bad, MF_2 scores 1.0, 0.1, or 100, respectively. Based on these proportions, the final MF_2 factor is evaluated. Hence, the tailored frequency is calculated as follows:

$$F = F_0 \times MF_2 \tag{6.6}$$

This method appears to be less static than API 581, but the technical aspects are not explicitly considered.

Finally, the barrier method [17] is based on indicators of safety barriers aimed at reducing the frequency of failures. This methodology requires gathering specific information related to safety barriers and is explicitly devoted to the assessment of technical issues with no direct consideration of managerial factors. Even if the method is complex and detailed, a more focused analysis of operational and organizational aspects is a key issue in the perspective of dynamic risk assessment.

5. THE RISK BAROMETER TECHNIQUE

Since 2006, the Norwegian oil and gas industry has focused its attention on integration of people, organizations, work processes, and information technology to make more effective knowledge-based decisions. This development has been supported by the Center for Integrated Operations in the Petroleum Industry (IO Center), a research-based innovation center established with support of the Research Council of Norway and the sponsorship of major international oil companies, suppliers, and research institutions [21].

One of the results obtained by the IO Center has been the development of a technique for the assessment of risk on a real-time basis, which may be classified as class IV in Fig. 6.1: the risk barometer [14]. The main focus of this technique is represented by the safety barriers in an industrial system. Sklet [22] defines safety barriers as passive or active, physical, technical, or human/operational systems aiming to prevent, control, or mitigate undesired events or accidents. The risk barometer assesses the performance of safety barriers by means of specific sets of indicators and relates this to the overall risk picture for evaluation of possible risk fluctuation.

Results can be visualized and shared in different sites to provide important decision support across geographical and disciplinary borders. For instance, it would allow defining both daily planning on an oil and gas platform offshore and discussing medium-term maintenance plans with engineers onshore.

The process of calibration of the risk barometer for specific industrial applications is carried out by means of a series of workshops with subject matter experts. The first workshop allows defining potential accident scenarios and identifying the barriers involved. In the following workshop, sets of indicators describing the performance of the barriers are outlined. Indicators include THO factors to assess the different barrier systems, as defined by Sklet [22]. Specific approaches for the development of such indicators are used, such as the resilience-base early warning indicators method described in Table 6.1. The use of indicators aimed at the improvement of the organizational attributes of resilience [12] serves to increase the proactivity of risk analysis. However, the availability of real-time data to substantiate the indicators is an important practical limitation when defining the indicators (to avoid a system that relies too heavily on manual input and assessment) [14].

To aggregate the information expressed by the indicators and assess the performance of barrier systems, barrier functions, and plant areas, the indicators are quantitatively weighted and combined by means of weighted summations (Fig. 6.2). Weighting and quantification depend on input from subject matter experts. In the last workshop, validation with real data from the plant is advisable to refine the modeling. However, continuous control and improvement of the indicators and the related weights should be always carried out.

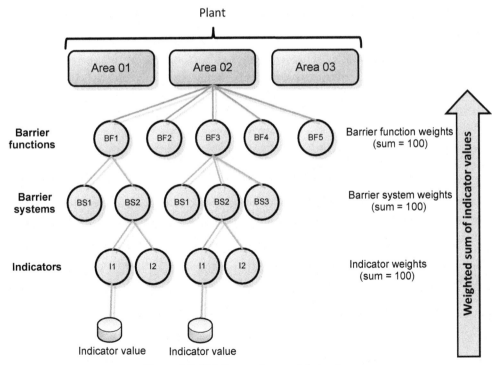

Figure 6.2 Risk Barometer: model structure.

5.1 Support to Decision-Making

As mentioned, the risk barometer is not exclusively a tool for dynamic risk assessment, but it was also developed to provide effective support to decision-making. For this reason, along the whole calibration process performed with subject matter experts, specific information is collected to be employed for improving decision support [23], such as:

- details on critical functions and strategies to maintain/restore them, identified through logic tree diagrams
- data needed to detect the occurrence of causes or consequences and the current health of the barriers
- key decisions made and the decision-makers who are designated to make these decisions

On the basis of this information, decision criteria are defined and related guidance messages are implemented in the risk barometer. Quantitative aggregation of the defined indicators allows evaluation of whether the decision criteria are satisfied, and the response guidance should be triggered and visualized by the tool. In this way the risk barometer not only monitors passive human and organizational factors but also contributes to their improvement.

6. CONCLUSIONS

This chapter introduces a series of proactive approaches for dynamic risk assessment. Such approaches are based on safety/risk indicators, which can be regularly collected and assessed to identify variations in the overall plant risk level. Moreover, human and organizational indicators can address underlying causes of unwanted events, and their analysis can enable proactive risk assessment. Four classes of approaches have been described, representing different levels of connection to the overall risk picture, ranging from techniques of safety indicator development, whose connection is assumed on the basis of past accident analysis, to dynamic aggregation methodologies based on comprehensive sets of THO indicators. In particular, a representative example of the latter has been presented, and its capability to support safety decision-making and prevent potential human and organizational failure has been described.

REFERENCES

[1] Paltrinieri N, Øien K, Cozzani V. Assessment and comparison of two early warning indicator methods in the perspective of prevention of atypical accident scenarios. Reliability Engineering & System Safety 2012;108:21−31.
[2] Johansen IL, Rausand M. Risk metrics: interpretation and choice. 2012. p. 1914−8.
[3] Health and Safety Executive. Developing process safety indicators. HSE guidance series, HSG254. Bootle, United Kingdom: HSE; 2006.

[4] Osborn RN, Olson J, Sommers PE, McLaughlin SD, Jackson JS, Scott WG, et al. Organizational analysis and safety for utilities with nuclear power plants. NUREG/CR-3215, PNL-4655, BHARC-400/83/011. An organizational overview, vol. 1. Washington, DC, USA: US Nuclear Regulatory Commission; 1983.

[5] Osborn RN, Olson J, Sommers PE, McLaughlin SD, Jackson MS, Nadel MV, et al. Organizational analysis and safety for utilities with nuclear power plants. NUREG/CR-3215, PNL-4655, BHARC-400/83/012. Perspectives for organizational assessment, vol. 2. Washington, DC, USA: US Nuclear Regulatory Commission; 1983.

[6] Boccio JL, Vesely WE, Azarm MA, Carbonaro JF, Usher JL, Oden N. Validation of risk-based performance indicators: safety system function trends. NUREG/CR-5323, BNL-NUREG-52186. Washington, DC, USA: US Nuclear Regulatory Commission; 1989.

[7] Wreathall J, Fragola J, Appignani P, Burlile G, Shen Y. UREG/CR-5436, SAIC- 90/1130. The development and evaluation of programmatic performance indicators associated with maintenance at nuclear power plants: main report, vol. 1. Washington, DC, USA: US Nuclear Regulatory Commission; 1990.

[8] Holmberg J, Laakso K, Lehtinen E, Johanson G. Safety evaluation by living probabilistic safety assessment and safety indicators. In: The Nordic Council of Ministers. Copenhagen, Denmark: TemaNord; 1994.

[9] Øien K, Sklet S. Application of risk analyses in the operating phase, establishment of safety indicators and modelling of organizational factors' effects on the risk level—a "state-of-the-art" description, STF38 A99416. Trondheim: Norway SINTEF Technology and Society, Safety Research; 1999.

[10] IAEA. Management of operational safety in nuclear power plant. INSAG-13. Vienna, Austria: International Nuclear Safety Advisory Group; 1999.

[11] Øien K, Utne IB, Herrera IA. Building safety indicators: part 1 — theoretical foundation. Safety Science 2011;49:148—61.

[12] Øien K, Utne IB, Tinmannsvik RK, Massaiu S. Building safety indicators: part 2 — application, practices and results. Safety Science 2011;49:162—71.

[13] Johnson JW, Rasmuson DM. The US NRC's accident sequence precursor program: an overview and development of a Bayesian approach to estimate core damage frequency using precursor information. Reliability Engineering & System Safety 1996;53:205—16.

[14] Hauge S, Okstad E, Paltrinieri N, Edwin N, Vatn J, Bodsberg L. Handbook for monitoring of barrier status and associated risk in the operational phase. SINTEF F27045. Trondheim: Norway Center for Integrated Operations in the Petroleum Industry; 2015.

[15] Center for Chemical Process Safety. Guidelines for chemical process quantitative risk analysis. New York, USA: America Institute of Chemical Engineers (AIChE); 2000.

[16] TNO. Guidelines for quantitative risk assessment: purple book. den Hague, The Netherlands: National Institute of Public Health and Environment (RIVM); 2005.

[17] Pitblado R, Bain B, Falck A, Litland K, Spitzenberger C. Frequency data and modification factors used in QRA studies. Journal of Loss Prevention in the Process Industries 2011;24:249—58.

[18] Beerens HI, Post JG. The use of generic failure frequencies in QRA: the quality and use of failure frequencies and how to bring them up-to-date. Journal of Hazardous Materials 2006;130:265—70.

[19] API Publication 581. Risk-based inspection base resource document. 2000.

[20] Pitblado RM, Williams JC, Slater DH. Quantitative assessment of process safety programs. Plant/ Operations Progress 1990;9:169—75.

[21] IO Center. Center for integrated operations in the petroleum industry. 2015 [accessed 01.12.15].

[22] Sklet S. Safety barriers: definition, classification, and performance. Journal of Loss Prevention in the Process Industries 2006;19:494—506.

[23] Paltrinieri N, Hauge S, Nelson WR. Dynamic barrier management: a case of sand erosion integrity. In: Podofillini L, Sudret B, Stojadinovic B, Zio E, Kröger W, editors. Safety and reliability of complex engineered systems: ESREL 2015. London, United Kingdom: Taylor & Francis Group; 2015.

CHAPTER 7

Reactive and Proactive Approaches: Tutorials and Example

G.E. Scarponi[1], N. Paltrinieri[2,3], F. Khan[4], V. Cozzani[1]
[1]University of Bologna, Bologna, Italy; [2]Norwegian University of Science and Technology (NTNU), Trondheim, Norway; [3]SINTEF Technology and Society, Trondheim, Norway; [4]Memorial University of Newfoundland, St John's, NL, Canada

1. INTRODUCTION

Two different approaches may be adopted for dynamic evaluation of accident frequency. Such approaches are generally based on either reactive or proactive assessment and may be represented by the Bayesian inference-based dynamic risk assessment (BIDRA) technique and the risk barometer technique, respectively.

BIDRA is a methodology for dynamic risk assessment based on Bayesian inference, and its objective is achieved by monitoring and processing data on incidents and near misses during the system's lifetime to refine failure probabilities (FPs) for safety barriers and consequently update potential accident frequency values.

The risk barometer is based on the definition and real-time monitoring of relevant indicators to continuously assess the health of safety barriers and evaluate their probability of failure. Such indicators monitor not only the technical performance of barriers, but also the associated operational and organizational systems. In this way, the risk barometer aims to capture early deviations within the organization, which may have the potential to facilitate barrier failure and accident occurrence.

This chapter describes step-by-step tutorials for the two approaches and a representative application to the same case study with the purpose of highlighting similarities and differences.

2. METHODOLOGY TUTORIAL

2.1 Bayesian Inference-based Dynamic Risk Assessment

The flowchart in Fig. 7.1 illustrates the steps for BIDRA.

2.1.1 Step 0: Scenario Identification

The application of hazard identification techniques, such as bow-tie and event tree analyses, is a prerequisite for BIDRA and allows identification of the potential accident scenarios to study. Specific chains of events, their interrelationships, and the related safety

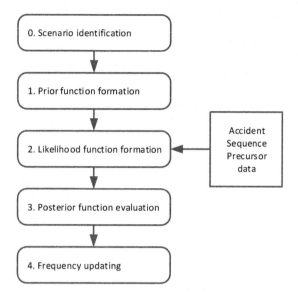

Figure 7.1 Bayesian inference-based dynamic risk assessment flowchart.

barriers can be outlined in a logic tree. In particular, definition of safety barriers (passive or active, physical, technical, operational, or human systems, or functions aiming at preventing, controlling, or mitigating undesired events or accidents [1]) is paramount. Knowledge of the reliability of each safety barrier is also needed and represents the starting point of the frequency updating process. Moreover, an effective system to collect and record incidents and near misses in which the safety barriers are involved [accident sequence precursor (ASP) data, representing data on failures and successes of barriers] must be implemented to drive such updating.

2.1.2 Step 1: Prior Function Formation

A prior failure function is identified for each safety barrier in the event tree/bow-tie diagram. This is a cumulative distribution function describing the FP of each and represents prior knowledge about safety barrier reliability before the operation is started. It is assumed that all the safety barrier failure probabilities follow the beta distribution. The discrete value of the safety barrier FPs can be expressed by the mean (μ) of the distribution (7.1), where α and β are the parameters by which the beta distribution is defined.

$$\text{FP} = \mu = \frac{\alpha}{\alpha + \beta} \tag{7.1}$$

If the discrete value of FP is the only available piece of knowledge on safety barrier reliability, the level of confidence in it, ie, the measure of the distribution dispersion around it (FP $= \mu$), should be defined through the distribution variance:

$$\text{VAR} = \frac{\alpha\beta}{(\alpha + \beta)^2(\alpha + \beta + 1)} \tag{7.2}$$

The higher the variance, the lower the degree of confidence, which also leads to higher sensitivity to posterior data in the following updates.

2.1.3 Step 2: Likelihood Function Formation

The likelihood function is a discrete function generated on the basis of the data collected about the performance of safety barriers, represented by regular tests, incidents, and near misses that occurred during the system lifetime (ASPs). In particular, when an unwanted event occurs, there may be both safety barriers that have failed and safety barriers that have succeeded in avoiding worse consequences. Therefore, it is important that such events be not only reported, but also investigated to recognize barrier failures and successes. The likelihood function is then defined by the set of failures and successes of the safety barriers studied. A convenient form to mathematically express the likelihood function is binomial distribution.

2.1.4 Step 3: Posterior Function Evaluation

If the prior function is in the form of beta distribution and the likelihood function is a binomial distribution, the posterior function obtained from them by means of Bayesian inference is in turn a beta distribution. For this reason, the updated FP of each safety barrier is represented by μ_1 of the posterior beta distribution:

$$\mu_1 = \frac{\alpha_1}{\alpha_1 + \beta_1} \tag{7.3}$$

$$\alpha_1 = \alpha_0 + F \tag{7.4}$$

$$\beta_1 = \beta_0 + S \tag{7.5}$$

where F equals the number of failures the safety barrier has encountered during the system lifetime and S equals the number of successes the safety barrier has encountered during the system lifetime.

Subscripts 0 and 1 identify the distribution parameters of the prior and the posterior functions, respectively.

2.1.5 Step 4: Frequency Updating

In the last step, the new FP of the safety barriers is taken into account for accident scenario frequency updating. Extrapolation of previous values of accident scenario frequency may allow for prediction of the future trend [2].

2.2 Risk Barometer

The flowchart in Fig. 7.2 illustrates the steps for the risk barometer method [3]. This is an iterative process, as indicated in the figure.

Figure 7.2 Risk barometer flowchart.

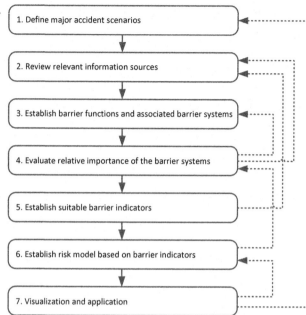

2.2.1 Step 1: Define Major Accident Scenarios
===

2.2.1 Step 1: Define Major Accident Scenarios

Step 1 aims to define the major accident scenarios to include in the risk barometer. The definition of the accident scenarios should also be conducted in cooperation with operational personnel. For this reason, a dedicated workshop can be conducted for this task. This selection should be based on the following two main criteria:

1. The selected event(s) must have a significant contribution to the major accident risk in terms of risk to personnel and/or environmental risk and/or loss of main safety functions.

2. Real-time information about related safety barriers can be collected and made available.

2.2.2 Step 2: Review Relevant Information Sources

The aim of step 2 is to identify and review relevant information sources. No single source of information is likely to provide all the required information. It is therefore necessary to combine various written information and information gained through discussions with operational personnel and subject matter experts.

Typical input sources include the following:

- quantitative risk assessment (QRA) of the installation under consideration
- qualitative and semiquantitative barrier analyses
- event reports and accident investigation reports
- various qualitative safety analyses

- selected system documentation and relevant procedures
- interviews, discussions, and workshops with experts/operational personnel

2.2.3 Step 3: Establish Barrier Functions and Barrier Systems

Step 3 aims to identify the barrier functions[1] and associated barrier systems[2] related to the defined major accident scenarios. Identification of barrier functions and barrier systems is generally based on QRA, experience and knowledge about the specific accident scenario, and preventive and mitigating measures in place to reduce the risk associated with the scenario. Use of logic trees (fault tree, event tree, or bow-tie diagram) from QRA, or specifically defined for this task, is advisable to preliminarily structure the following modeling.

2.2.4 Step 4: Evaluate Relative Importance of the Barrier Systems

The aim of step 4 is to evaluate the relative importance of the safety barriers to facilitate a risk-based selection of indicators. Sensitivity analysis is performed on the barrier i by assessing its Birnbaum-like measure $I^B(i)$ (based on Rausand [4]). This represents the partial derivative of the risk measure R with respect to the parameter p_i, which, in turn, describes the barrier and generally coincides with the barrier FP:

$$I^B(i|t) = \frac{\partial R(t)}{\partial p_i(t)} \tag{7.6}$$

Thus, risk at time t is defined as:

$$R(t) = R_0 + \sum_i I^B(i|t) \cdot \Delta p_i(t) = R_0 + \Delta R(t) \tag{7.7}$$

where R_0 equals the value of risk at a reference time 0 (eg, the time in which the QRA is performed); $\Delta p_i = p_{i,t} - p_{i,0}$; $p_{i,t}$ equals the value of the parameter i at time t; and $p_{i,0}$ equals the value of the parameter i at a reference time 0 (eg, the time in which the QRA is performed).

Reiteration of steps 2 and 3 may help to collect more details on the barrier systems and/or improve their definition.

2.2.5 Step 5: Establish Suitable Barrier Indicators

To establish suitable barrier indicators, barrier requirements (summarized into the barrier function) and factors/conditions influencing the overall risk [risk influencing factors

---

[1] A barrier function is the task or role of a barrier. Examples include preventing leaks or ignition, reducing fire loads, ensuring acceptable evacuation, and preventing hearing damage.

[2] The terms *barrier* and *barrier system* are often used interchangeably. A barrier system is typically a collection of barrier elements, such as technical, operational, or organizational measures or solutions that play a part in realizing a barrier function.

(RIFs)] should be taken into account. The set of indicators should be grouped into RIFs and should monitor the fulfillment of barrier requirements. Priority is given to the definition of indicators for the most risk-affecting barriers, as assessed in step 4. Discussions and reviews together with operational personnel are paramount both to define indicators and to ensure that data and information sources to support the indicators are available during operation. (For this reason step 2 may be reiterated.)

The four main criteria to consider for selection of indicators are the following:

1. the importance of the associated safety barrier
2. the relationship between the status of the indicator and the probability of failure of the associated barrier
3. the possibility of obtaining an objective measure of the indicator
4. the availability of real-time data to substantiate the indicator

2.2.6 Step 6: Establish Risk Model Based on Barrier Indicators

Step 6 addresses the question of how to aggregate individual barrier status information to obtain an overall risk measure. The model can be based on a selection of the safety barriers that most affect the overall risk (as assessed in step 4). This selection of barriers is at the analyst's discretion, and step 4 can be reiterated to refine the assessment performed. Once the barriers to consider in the model are identified and the related set of indicators are grouped into RIFs, the aggregation rules described in Table 7.1 should be followed.

2.2.7 Step 7: Visualization and Application

In step 7 the format for visualization of the risk picture should be established for area level, as well as for barrier function and barriers system performance. Typical visualization formats include the use of traffic light systems, radar charts, time-series trends, and two-dimensional criticality plots. Furthermore, the structure for drill-down capabilities should be determined. Drill-down capability provides higher levels of detail and better understanding of underlying conditions by showing detailed data.

Visualization provides decision support by presenting and monitoring the current risk and trends in the risk picture. Typical elements to visualize are the following:

- total risk level with trend
- risk level for a specific location
- top five risk contributors, taking, for example, current values and trends into account
- risk contribution for a set of safety barriers
- safety barrier performance over time

Further details on risk visualization may be found in Chapter 9 and elsewhere [3].

Table 7.1 Risk Barometer Aggregation Rules

Model Level	Aggregation Rules
Indicator	The value of each indicator (x_i) should be translated into mutually comparable scores from 1 to 6, where 1 is the most positive score and 6 the most negative. Such translation may be obtained by means of direct proportionality, inverse proportionality, or, if needed, other mathematical functions at the analyst's discretion. $$s_i = S(x_i) \qquad (7.8)$$ A percentage weight (w_i) is assigned to each indicator i reflecting its relative importance (in relation to the other indicators) in the definition of a score describing the status of the upper level. The upper level may be represented by the barrier system or the associated risk influencing factor (an intermediate level used to group indicators, which is omitted from this description but follows the same aggregation rule of weighted summation).
Barrier system	A score describing the status of the barrier system j (BS_j) is obtained on a scale from 1 to 6 (1 = most positive, 6 = most negative) by means of weighted summation. $$BS_j = \sum W_i S_i \qquad (7.9)$$ A percentage weight (w_j) is assigned to each barrier system j reflecting its relative importance (in relation to the other barrier systems) for the definition of a score describing the status of the barrier function.
Barrier function	A score describing the status of the barrier function k (BF_k) is obtained on a scale from 1 to 6 (1 = most positive, 6 = most negative) by means of weighted summation. $$BF = \sum W_j BS_j \qquad (7.10)$$
Failure probability	The score describing the status of the barrier function is translated into a probability of failure (p). Such translation may be obtained by means of direct proportionality with the realistic variation range of failure probability defined in step 4. Other mathematical functions may be also applied at the analyst's discretion. $$p = P(BF) \qquad (7.11)$$

3. APPLICATION OF REACTIVE AND PROACTIVE APPROACHES

An example of application of both the reactive and proactive approaches, represented by BIDRA and the risk barometer, respectively, is performed on the topside process of an oil and gas offshore platform. A preliminary risk assessment exercise representing the QRA of the platform is used as a basis for the application of the two approaches.

3.1 Topside Process of an Oil and Gas Offshore Platform

The process can be divided into four modules (Fig. 7.3): (1) gas compression and recompression, (2) choke and manifold, (3) separation and test, and (4) water treatment.

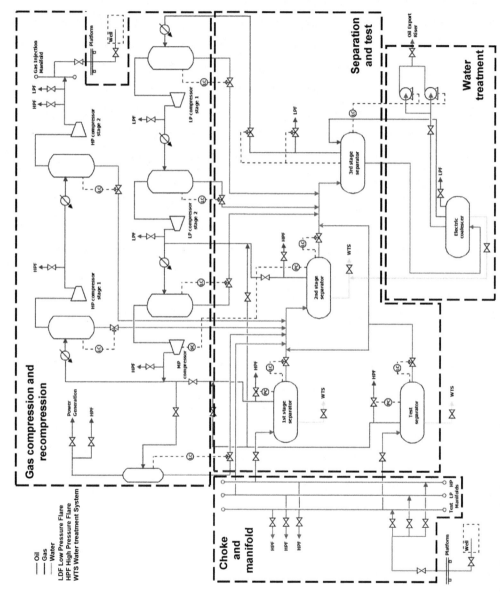

Figure 7.3 Topside process divided into modules.

The preliminary risk quantification process performed for each module may be summarized in four steps. (Further details can be found elsewhere [2,5].)

1. Application of MIMAH (methodology for the identification of major accident hazards) [6], allowing for hazard identification. Fig. 7.4 reports the event trees defined for liquid leak and gas leak in the separation and test module. Such event

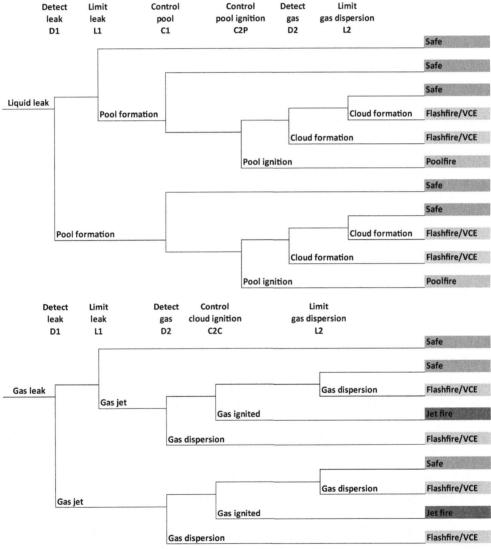

Figure 7.4 Event trees for liquid and gas leaks obtained by means of the methodology for the identification of major accident hazards (VCE Stands for Vapour Cloud Explosion) [6].

Table 7.2 Safety Barriers and Relative Failure Probabilities as Defined by the Methodology for the Identification of Major Accident Hazards and the Method for Identification of Reference Accident Scenarios (to Clarify the Defined Safety Barriers, the Main Associated Device Is Indicated) [6]

Label	Safety Barrier	Failure Probability
D1	Detect leak (low pressure alarm)	0.100
L1	Limit leak (emergency valve)	0.024
C1	Control pool (sump system)	0.500
C2P	Control pool ignition	0.010
C2C	Control cloud ignition	0.015
D2	Detect gas (gas detector)	0.014
L2	Limit gas dispersion (water spray barrier)	0.013

trees can also be used for the modules of choke/manifold and water treatment, whereas only the gas leak event tree can be applied to the module of gas compression and recompression.

2. Application of the method for identification of reference accident scenarios [6], allowing for evaluation of FP of the safety barriers and frequency of the identified potential accident events. Table 7.2 reports the identified safety barriers and the related failure probabilities, whereas Table 7.3 summarizes the frequencies of the potential accident events.

3. Classification of the event typologies and the related potential impact by means of consequence classes defined by Kalantarnia et al. [2], allowing for impact evaluation of the identified potential events. Table 7.3 summarizes the impact of consequences expressed as equivalent economic loss.

4. Assessment of risk related to the potential accident events and the plant modules expressed as equivalent economic loss per year (see Table 7.3).

3.2 Bayesian Inference-based Dynamic Risk Assessment

The preliminary risk assessment performed for the case fulfills the prerequirements for the application of BIDRA.

In the first step, the prior function for each safety barrier is identified. To outline a beta distribution representing the prior function, the parameters α and β should be identified. In this case the parameters are not provided, and only a discrete value of FP is available. For this reason, α and β should be selected by means of Eqs. (7.1) and (7.2). In fact, there are infinite pairs of parameters for which Eq. (7.1) is satisfied. To univocally determine the values of α and β, it is necessary to define the variance of the distribution and

Table 7.3 Summary of the Results Provided by the Preliminary Risk Assessment [6]

Module		Flash Fire/Vapor Cloud Explosion (Liquid)	Pool Fire	Flash Fire/Vapor Cloud Explosion (Gas)	Jet Fire	Risk ($/year)
Gas compression and recompression	Frequency (events/year)			1.95×10^{-5}	1.08×10^{-5}	1.03×10^{4}
	Loss ($)			5.00×10^{8}	5.00×10^{7}	
	Risk ($/year)			9.77×10^{3}	5.40×10^{2}	
Choke/manifold	Frequency (events/year)	1.62×10^{-5}	6.08×10^{-6}			2.84×10^{3}
	Loss ($)	1.00×10^{8}	2.00×10^{8}			
	Risk ($/year)	1.62×10^{3}	1.22×10^{3}			
Separation and test	Frequency (events/year)	4.38×10^{-6}	1.64×10^{-6}	8.79×10^{-6}	4.86×10^{-6}	5.05×10^{3}
	Loss ($)	1.00×10^{8}	1.00×10^{8}	5.00×10^{8}	1.00×10^{7}	
	Risk ($/year)	4.38×10^{2}	1.64×10^{2}	4.40×10^{3}	4.86×10^{1}	
Water treatment	Frequency (events/year)			3.26×10^{-7}	1.80×10^{-7}	2.13
	Loss ($)			1.00×10^{6}	1.00×10^{7}	
	Risk ($/year)			3.26×10^{-1}	1.80×10^{0}	

satisfy Eq. (7.2). Thus, the equations defining the prior values of α and β are the following:

$$\beta_0 = \left(\frac{FP \times (1 - FP)}{VAR} - 1 \right) \times (1 - FP) \tag{7.12}$$

$$\alpha_0 = \frac{FP \times \beta_0}{(1 - FP)} \tag{7.13}$$

where α_0 and β_0 are the values of the parameters in the year 0.

A first tentative value of variance equal to 10^{-4} is used in this example, which is later demonstrated by a sensitivity analysis performed on the posterior function (Fig. 7.5). Table 7.4 reports the calculated prior values of α and β of each safety barrier in Fig. 7.4 in the year 0.

In the second step, the formation of the likelihood function is required. Comprehensive monitoring of the performance of safety barriers is assumed for this step. Table 7.5 reports the number of failures and successes in a cumulative form over a time span of 10 years (considering regular tests, incidents, and near misses) for each safety barrier represented in Fig. 7.4.

Once both prior and likelihood functions are available, the posterior function is calculated (step 3). For the year i and the barrier j, the distribution parameters $\alpha_{i,j}$

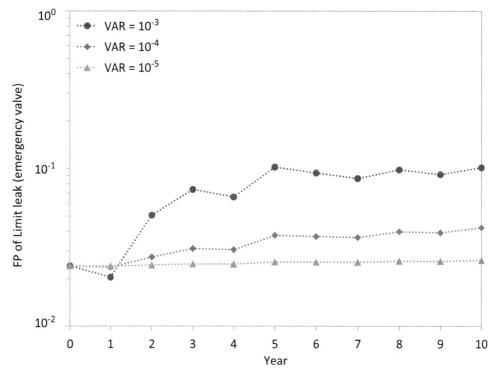

Figure 7.5 Representative example of beta distribution variance (VAR) sensitivity analysis applied on the barrier "limit leak (emergency valve)."

Table 7.4 Prior Values of α and β in the Year 0 for Each Safety Barrier Identified

Safety Barrier	α_0	β_0
Detect leak (low pressure alarm) — D1	89.9	809
Limit leak (emergency valve) — L1	5.64	228
Control pool (sump system) — C1	1250	1250
Control pool ignition — C2P	0.98	97
Control cloud ignition — C2C	2.2	145
Detect gas (gas detector) — D2	1.92	135
Limit gas dispersion (water spray barrier) — L2	1.69	127

Table 7.5 Number of Failures and Successes in a Cumulative Form Over a Time Span of 10 years for Each Safety Barrier Identified

	Safety Barrier												
	D1		L1		C1		C2P		C2C		D2		L2	
Year	F	S	F	S	F	S	F	S	F	S	F	S	F	S
1	0	4	0	4	1	3	0	4	1	3	0	4	1	3
2	2	7	1	7	3	6	0	9	1	7	0	9	2	7
3	2	12	2	11	4	9	1	12	2	10	1	12	3	10
4	2	15	2	14	4	13	1	16	2	14	1	16	4	13
5	3	19	4	17	6	16	1	21	3	17	1	21	5	17
6	3	23	4	21	7	19	2	24	4	20	2	24	5	21
7	4	26	4	25	8	22	3	27	5	23	2	28	6	24
8	5	30	5	28	9	26	3	31	5	27	3	31	6	28
9	5	34	5	32	9	30	3	35	5	32	4	34	6	32
10	6	38	6	36	10	33	3	39	5	36	4	38	7	35

F, failure; S, success.

and $\beta_{i,j}$ are obtained using Eqs. (7.4) and (7.5). Eq.(7.3) allows calculation of the updated $FP_{i,j}$.

These first three steps can be repeated considering different values of variance to perform a sensitivity analysis and assess whether the level of FP sensitivity to posterior data is appropriate. Such sensitivity analysis is applied on the barrier "limit leak (emergency valve)" as a representative example (Fig. 7.5).

On the basis of the updated FP of safety barriers, the occurrence frequencies of the accident events identified in Fig. 7.4 are calculated for each year (step 4). Fig. 7.6 illustrates the trend of the updated accident event frequencies (the frequencies of flash fire/vapor cloud explosion due to liquid and gas leakage are merged together) over the time span of 10 years.

3.3 Risk Barometer

The major accident scenarios to include in the risk barometer (step 1) are the ones mentioned in the preliminary risk assessment (and represented in Fig. 7.4) of the considered topside process of an oil and gas offshore platform.

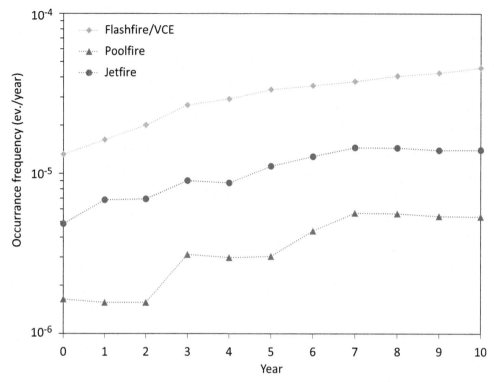

Figure 7.6 Bayesian inference-based dynamic risk assessment of accident event frequency trends over a time span of 10 years.

Such preliminary risk assessment also represents the main information source (step 2) for the application of the risk barometer approach.

Moreover, the safety barriers reported in the event trees in Fig. 7.4 are used for the identification of barrier functions and barrier systems (step 3). For each barrier function, only one safety barrier is initially identified in the preliminary risk assessment. (In some cases, a technical element is also mentioned for exemplificative purposes.) For this reason, safety functions and safety systems are merged together and generically denominated as "safety barriers" (see Table 7.2).

The next step (step 4) consists of the evaluation of the relative importance of safety barriers. Barrier FPs are considered as the parameters of interest (p) in the assessment of the Birnbaum-like measure (Eq. (7.6)). In this example, a specific value for the Birnbaum-like measures is assessed for each safety barrier FP at a time (Table 7.6) on the basis of the following:

- Δp = realistic ranges in which the FP values may vary, defined on the basis of expert judgment
- ΔR = associated risk variation with respect to the initial risk R_0, calculated by means of the risk model used in the preliminary risk assessment study

Table 7.6 Sensitivity Analysis Parameters and Results; Gray Cells Highlight a Risk Variation Lower Than 5%

Barrier Label	Birnbaum-like measure I^B	Realistic Failure Probability Variation Range $\pm \Delta p$ %	Risk Variation $\pm \Delta R$ %
Choke/manifold			$R_0 = 2840.43 \text{ \$/year}$
D1	2.28×10^4	50%	40.10%
L1	2.10×10^4	82%	14.61%
C1	5.68×10^3	80%	80.00%
C2P	1.20×10^5	85%	35.92%
D2	5.94×10^4	67%	19.63%
L2	5.94×10^4	67%	18.41%
Separation and test			$R_0 = 5047.36 \text{ \$/year}$
D1	4.05×10^4	50%	40.10%
L1	3.73×10^4	82%	14.61%
C1	1.21×10^3	80%	9.55%
C2P	1.60×10^4	85%	2.69%
C2C	1.11×10^3	64%	0.21%
D2	1.78×10^5	67%	33.11%
L2	1.76×10^5	67%	30.62%
Gas compression/recompression			$R_0 = 10309.07 \text{ \$/year}$
D1	8.27×10^4	50%	40.10%
L1	7.62×10^4	82%	14.61%
C2C	3.13×10^4	64%	2.91%
D2	3.60×10^5	67%	32.73%
L2	3.55×10^5	67%	30.28%
Water treatment			$R_0 = 2.13 \text{ \$/year}$
D1	1.70×10^1	50%	40.10%
L1	1.57×10^1	82%	14.61%
C2C	1.20×10^2	64%	54.12%
D2	1.02×10^1	67%	4.50%
L2	1.18×10^1	67%	4.90%

A threshold of risk variation equal to 5% is set to take into account the most risk-affecting barriers. Safety barriers leading to a potential risk variation lower than 5% (gray cells in Table 7.6) are disregarded by the following methodology steps.

Barrier indicators are defined on the basis of the report "Guideline for implementing the REWI method" [7] (step 5). As a representative example, Table 7.7 reports the indicators describing the status of the safety barrier "limit leak L1" and the related grouping under the RIFs.

Indicator and RIF weights are defined on the basis of expert judgment (see Table 7.7) and are used for the definition of a risk model following the aggregation rules reported in Table 7.1 (step 6). In this example the risk model includes the following levels: (1) indicators, (2) RIFs, (3) safety barriers, and (4) FP.

Table 7.7 Indicators and Status of the Safety Barrier "Limit Leak L1"

Safety Barrier	Risk Influencing Factor	Risk Influencing Factor Weight (%)	Indicator	Indicator Weight (%)
Limit leak (emergency valve) L1	Performance	50	Number of failures on demand	66
			Number of failed tests	34
	Operational support	25	Number of persons responsible for monitoring the related control panel	100
	Maintenance	25	Number of inspections/ audits performed	35
			Number of functional tests performed	35
			Portion of maintenance personnel receiving training	30

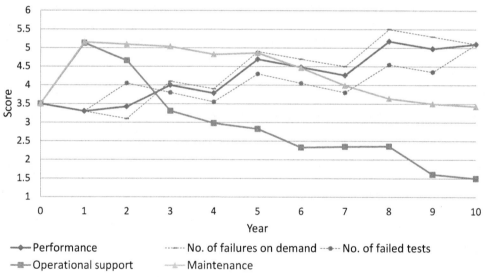

Figure 7.7 Indicators and risk influencing factors of the safety barrier "limit leak (emergency valve)" over a time span of 10 years.

The risk model obtained is applied on sets of indicator values simulated for a time span of 10 years. Fig. 7.7 reports the value of two representative indicators (number of failures on demand and number of failed tests, reflecting the data from Table 7.5). Such indicators allow definition of the status of the RIF "performance," which tends to worsen and increase in score over a period of 10 years. On the contrary, both the RIFs of "operational support" and "maintenance" have an initial increase in score, followed by a general improvement and score decrease. This simulates the organization's reaction after the occurrence of technical failures. However, despite the delay such improvements in operational support and maintenance may cause before resulting into a steady decrease of the performance score, their trends show a healthy response of the

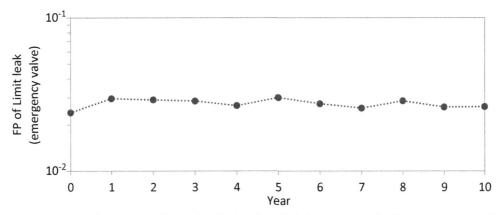

Figure 7.8 Failure probability of the safety barrier "limit leak (emergency valve)" over a time span of 10 years.

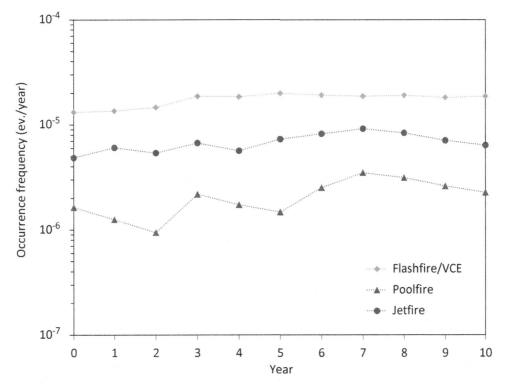

Figure 7.9 Risk barometer accident event frequency trends over a time span of 10 years.

system. For this reason, the FP of the barrier under study tends to assume a relatively flat trend over time (Fig. 7.8).

In general, the risk barometer considers more information related not only to the technical performance of the safety barriers but also to the operational and organizational aspects supporting them. For this reason, the overall frequency trends of the accident events differ from the BIDRA results and lead to a less critical increase over time (Fig. 7.9).

The risk barometer technique also aims to visualize the current overall risk and its variation over time (step 7). Such visualization capabilities are not addressed in this chapter, but they are described in detail in Chapter 9.

4. CONCLUSIONS

This chapter illustrates two tutorials for the application of the BIDRA and risk barometer techniques, which are reactive and proactive approaches, respectively, for dynamic frequency analysis. Whereas BIDRA allows for updating accident event frequency on the basis of past events, the risk barometer aims to also monitor the health of an organization to obtain a comprehensive picture of the system and capture early deviations possibly leading to unwanted events.

Records of near misses and incidents represent the inputs for BIDRA, whereas real-time technical, operational, and organizational indicators are used by the risk barometer to evaluate the overall risk variation over time. Such information has been simulated in this chapter for the application of the two techniques on the same representative case. This exercise has showed their similarities and differences and may provide support for the choice between reactive and proactive approaches of dynamic risk analysis.

REFERENCES

[1] Sklet S. Safety barriers: definition, classification, and performance. Journal of Loss Prevention in the Process Industries 2006;19:494—506.
[2] Kalantarnia M, Khan F, Hawboldt K. Modelling of BP Texas City refinery accident using dynamic risk assessment approach. Process Safety and Environmental Protection 2010;88:191—9.
[3] Hauge S, Okstad E, Paltrinieri N, Edwin N, Vatn J, Bodsberg L. Handbook for monitoring of barrier status and associated risk in the operational phase. SINTEF F27045. Trondheim: Norway Center for Integrated Operations in the Petroleum Industry; 2015.
[4] Rausand M. Risk assessment: theory, methods, and applications. Wiley; 2013.
[5] Paltrinieri N, Scarponi GE, Khan F, Hauge S. Addressing dynamic risk in the petroleum industry by means of innovative analysis solutions. Chemical Engineering Transactions 2014:451—6.
[6] Delvosalle C, Fievez C, Pipart A, Debray B. ARAMIS project: a comprehensive methodology for the identification of reference accident scenarios in process industries. Journal of Hazardous Materials 2006; 130:200—19.
[7] Øien K, Massaiu S, Ranveig KT. Guideline for implementing the REWI method. Trondheim, Norway: SINTEF; 2012.

CHAPTER 8

Comparison and Complementarity between Reactive and Proactive Approaches

G.E. Scarponi[1], N. Paltrinieri[2,3]
[1]University of Bologna, Bologna, Italy; [2]Norwegian University of Science and Technology (NTNU), Trondheim, Norway; [3]SINTEF Technology and Society, Trondheim, Norway

1. INTRODUCTION

The previous chapters have introduced two types of approaches for dynamic risk analysis: the Bayesian inference-based dynamic risk assessment (BIDRA) [1,2], which may be defined as reactive, and the risk barometer [3], which may be defined as proactive.

BIDRA monitors and processes data on incidents and near misses that have occurred during the system's lifetime. It is based on Bayesian inference, which allows for updating of failure probabilities of safety barriers (and related accident frequency values) on the basis of the information collected. On the other hand, the risk barometer aims to continuously assess health of safety barriers and their probability of failure on the basis of specific indicators. Such indicators cover not only the technical performance of a barrier system but also the associated operational and organizational contributors to risk.

These approaches aim to overcome the inherent limitations resulting from the static nature of conventional risk assessment techniques [4]. However, they address this issue from different perspectives, which result in both benefits and limitations. This chapter addresses such differences in reactive and proactive perspectives by describing the distinctive traits of BIDRA and the risk barometer. Moreover, a qualitative comparison is discussed, and remarks on potential complementarity are drawn.

2. BENEFITS AND LIMITATIONS OF REACTIVE AND PROACTIVE APPROACHES

The ISO 31000 standard on risk management [5] assigns a pivotal role to knowledge by defining risk as "the effect of uncertainty on objectives." Uncertainty is the driving force of dynamic risk analysis, demanding continuous calibration of the risk picture and progressively replacing lack of knowledge with new evidence and information. Moreover, awareness of the knowledge dimension, as theorized by Aven [6] and Aven and Krohn [7], not only gives credit to iterative risk assessment but also improves its understanding.

Dynamic Risk Analysis in the Chemical and Petroleum Industry
ISBN 978-0-12-803765-2

Within the framework outlined by this book, risk analysis is defined as *dynamic* if it is capable of continuously iterating, updating, and keeping track of changing system conditions. However, knowledge of *when* data are collected and risk is assessed is essential to understand the limits of such an approach.

In addition, the distinction between reactive and proactive approaches reflects different projections in time for the risk assessed. Reactive approaches respond to an event that is directly associated with the overall risk picture (eg, failure or success of a technical safety system) and is presumably closer in time to a potential accident, if not to an accident itself. Proactive approaches, however, also include in the analysis relevant early deviations from the optimal condition, which have a lower degree of causality on a potential accident (eg, worsening or improvement of organizational factors).

Fig. 8.1 represents the distinction between reactive and proactive approaches for dynamic risk analysis. Performance of a hypothetical industrial system (including both plant and organization) is plotted against time in blue. Such performance is defined for representative purposes only and attempts to summarize the overall system condition. It is inspired by the company overview of environmental, social, and governance (ESG) performance provided by Thomson Reuters Eikon [8]. However, like the ESG performance, this performance does not reflect the overall risk, which is represented by the two dotted red lines.

Worsening of system performance may be due to several factors, among which there may be risk contributors: technical failures or deterioration of safety equipment as well as organizational inefficiency or human inexperience. Improvement from the optimal system performance is assumed possible, but not sustainable for a long period. In Fig. 8.1, events of performance degradation are followed by corrective responses, which, for explanatory purposes, are assumed to be focused on maintaining the designed production and relatively disregarding safety implications.

In fact, Fig. 8.1 reproduces the occurrence of an accident and the previous warning signals that may be appreciable on the system performance. Reactive approaches would account for safety equipment failures and related corrective actions and would dynamically adjust the overall risk assessment after each event—accident included. Proactive approaches would behave in the same fashion but would account for earlier events, such as potential organizational and operational issues (lack of training and maintenance backlog). This gives different time perspectives to the approaches, which should be known and understood for correct use.

2.1 Bayesian Inference-Based Dynamic Risk Assessment

The BIDRA technique is based on sound statistical theories and falls under the definition of reactive approaches [1,2]. In fact, it updates the risk picture of the system by considering information on past events that indicated failure or success of safety barriers. For instance, the example of application proposed in the previous chapter shows how this technique can identify worsening in the safety system or a negative drift toward risk conditions through the registered failures in regular tests of safety-instrumented

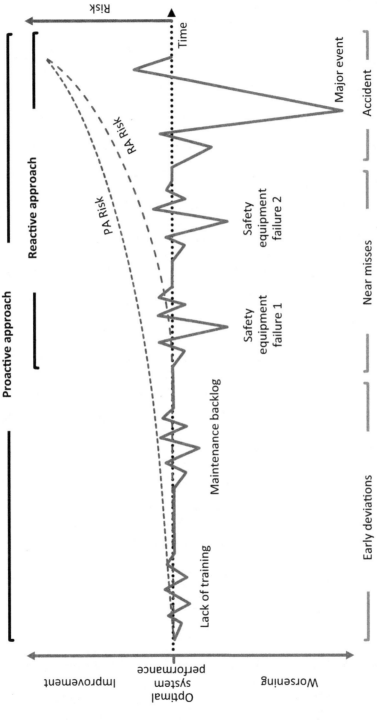

Figure 8.1 Distinction between reactive and proactive approaches for dynamic risk analysis. *PA risk*, Risk assessed by means of a proactive approach; *RA risk*, risk assessed by means of a reactive approach.

systems. However, the definition of success and failure of a safety barrier poses some limitations on this technique.

Classification of an event into success or failure may be challenging, and it is not described in detail by the literature [1,2]. The approach defined by the United Kingdom's Health and Safety Executive to define lagging indicators may be an inspiration [9]. A desired outcome for a safety barrier should be defined, eg, the desired function the barrier is designed to deliver. Success is represented by the achievement of such an outcome; otherwise it is a failure. No intermediate conditions are allowed by the technique.

Because of the specific characteristics of the data used as input to BIDRA, technical information on the performance of safety equipment is relatively more suitable. This performance may affect the probability of an accident to a higher degree than operational and organizational factors because it is closer in the causality chain. For this reason, the hypothetical representation given by Fig. 8.1 shows such failures relatively close in time to the accident and the risk assessed by a reactive approach suddenly increasing once they occur. However, not only incidents and near misses but also results of regular technical tests can be input to BIDRA, which would allow constant updating with known lag. The main requirement for BIDRA inputs is a certain degree of objectivity, allowing for distinction between success and failure of safety barriers and providing a relevant and critical basis for the analysis.

BIDRA is presented in the previous chapter as an integrative technique that is able to correct and calibrate prior analyses based on generic failure probabilities of safety barriers and to update them to reflect the real system condition. Step 0 of the tutorial in Chapter 7 represents the prerequired initial analysis (despite the fact that BIDRA may be also described as a stand-alone technique [1]). This shows a direct link with the analysis from the design phase, which is regulated by the variance used. Such variance expresses the sensitivity of barrier failure probabilities to posterior data and may be defined *a priori* by the beta distribution parameters. If no parameters are available, the level of confidence in the initial assessment (and the following updates) should be defined to allow for more (or less) radical posterior adjustments. Sensitivity analysis may be of support in this regard (Fig. 8.2).

The adopted variance is the expression of the knowledge dimension mentioned in the previous section. Despite the solid statistical foundation and the relevant data input, there is room for subjectivity in the application of BIDRA. Experience-based decisions may be made to define a variance value for the dynamic analysis. For this reason, the related uncertainties should be clearly highlighted during the use of BIDRA to raise the analyst's awareness and search for appropriate support.

2.2 Risk Barometer

The risk barometer is a relatively recent technique [3], which may be defined as proactive. It is based on a wide range of indicators monitoring technical, operational, and organizational factors affecting the overall risk. Deviations of indicators are reflected in the

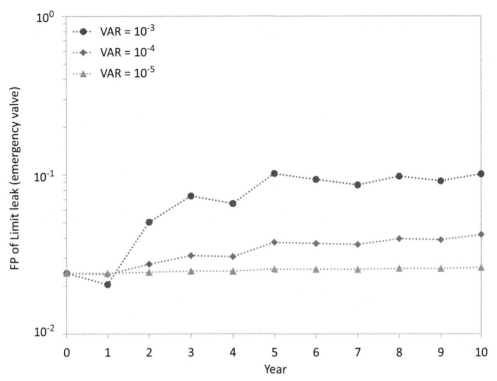

Figure 8.2 Representative example of variance (VAR) sensitivity analysis carried out for the yearly update of a safety barrier failure probability.

failure probability of safety barriers and, in turn, in the risk picture by means of specific aggregation rules. For instance, the previous chapter reports how the method can assess risk variations from the system's technical performance and the related organization's health in a simulated time span. However, the evaluation of risk contribution by the system's safety barriers, the definition of related indicators, and the aggregation of the information they convey are the main limitations of the approach.

The risk barometer is presented in Chapter 7 as a stand-alone technique for which no specific prerequirement seems to be needed. (No step 0 is reported in the risk barometer tutorial in Chapter 7.) This is suggested by the fact that the risk barometer is relatively adaptable to the degree of available information present for the case. Application of the risk barometer to a variety of safety-critical cases, such as scenarios of oil leakage, sand erosion, and ship collision, is reported elsewhere [3]. However, the evaluation of the relative importance of the system's safety barriers (and the omission of parameters representing the less influential safety barriers) should be preferably performed on the basis of previous risk and barrier analyses. In case pieces of information are not available, the evaluation may be based on expert judgment and subject to a higher level of uncertainty. Poor judgment may result in the exclusion of critical parameters.

Suitable barrier indicators are defined for the application of the risk barometer. The handbook describing the risk barometer technique [3] lists four main criteria for the development of suitable indicators (also reported in Chapter 7). However, further support may be found in additional reports:

- "Developing process safety indicators," by the Health and Safety Executive (United Kingdom) [9], mainly for the definition of technical and operational indicators.
- "Guideline for implementing the REWI method," by SINTEF (Norway) [10], mainly for the definition of operational and organizational indicators.

The risk barometer characteristic of proactivity resides in its capacity to also consider and process underlying operational and organizational factors, which may affect the performance of safety-instrumented systems during operations [11]. Such factors may be earlier in the causality chain than a technical failure and, for this reason, are represented as early deviations in the sequence of events leading to an accident (see Fig. 8.1, "lack of training" and "maintenance backlog"). The technique anticipates the evaluation of accident risk increase on the basis of such signals. For this reason, the risk assessed by a proactive approach (depicted in Fig. 8.1) starts increasing as soon as such deviations are detected.

The indicators collected are heterogeneous and should be translated into mutually comparable scores. Such translation may be obtained through direct proportionality, inverse proportionality, or other mathematical functions at the analyst's discretion (Fig. 8.3). Selection of the right function is highly critical because it affects the basis of dynamic risk analysis itself.

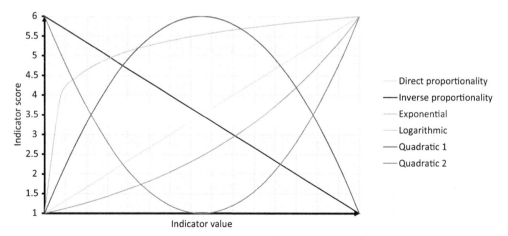

Figure 8.3 Examples of functions for translating indicator values into scores.

The different nature of the indicators collected by the risk barometer also affects the time lag with which risk is dynamically updated. Organizational, operational, and technical indicators have different collection frequencies, which may alter the overall result. In particular, organizational indicators may not be characterized by a regular interval of collection (eg, participation in trainings), whereas technical indicators are constantly produced (eg, automated tests). Such an issue has been addressed only qualitatively in the previous applications of the risk barometer, and it represents an important uncertainty for the analyst.

Organizational and operational indicators are valuable in predicting potential critical events and correcting deviations that may potentially develop in accidents. However, their link with the overall risk is not as direct as technical indicators, and they are relatively challenging to define (such challenges are addressed elsewhere [10]), which may lead to omission or double counting. For this reason, the risk barometer accounts for the relative importance of indicators (and other elements of the hierarchical structure, as shown in Chapter 7) in respect of the overall risk by means of specific weights and weighted summations. Such weights are defined on the basis of expert judgment. This may be time consuming and lead to new uncertainties. As suggested by the author [12], the technique of analytical hierarchy process [13] or the more recent measuring attractiveness by a categorical based evaluation technique [14] can be applied to facilitate the process.

Other methodologies, such as risk modeling—integration of organizational, human, and technical factors [15] and the organizational risk influence model [16,17], propose articulated structures to account for organizational, operational, and technical factors. However, the set-up of these techniques requires more effort and time and may not appeal in an industrial context. In this regard, the hierarchical structure of the risk barometer is easily understandable and allows for drill down capabilities. In fact, risk variation may be traceable up to the single indicators. Such transparency in the model is promoted as a specific characteristic of the risk barometer: the "transparent box philosophy," in contrast with the black box concept where only inputs and outputs can be known. Such an approach aims not only to build trust in the potential user but also to further address the knowledge dimension and raise awareness of the related uncertainties.

3. COMPARISON AND COMPLEMENTARITY OF THE TWO TECHNIQUES

3.1 Comparison

It is worth noting that the terms *reactive* and *proactive* do not rigidly apply to methods. In fact, BIDRA not only collects information from previous events (near misses or incidents) but also considers the results of tests of safety-instrumented systems, which provides a certain degree of proactivity in the analysis. On the other hand, the risk barometer elaborates both reactive and proactive indicators (as represented in Fig. 8.1), and its applications have proved to rely on the former to a greater extent [3]. For this reason, such classification may be reductive, and the overlap between the two techniques may be considerable in some cases.

Important differences between the techniques stand out concerning the risk updating process. BIDRA considers the components of the process at a rather superficial level. It may evaluate the current failure probability of a safety barrier on the basis of its behavior in the past. Such information may support maintenance planning and lead to corrective maintenance or risk management in general as well as additional safety barriers. However, the technique does not allow investigation of the possible underlying causes of malfunctioning. On the other hand, the risk barometer provides a deeper insight of the causes. It focuses on factors affecting the general behavior of the system. In this way, it makes possible the identification of a negative drift at an early stage of the cause-consequence chain leading to an undesired event.

3.2 Complementarity

Two specific remarks may be mentioned regarding the complementarity of the methods.

1. The risk barometer represents a further development of BIDRA. In fact, the risk barometer takes into account underlying factors (addressing organization health and operations) in addition to the test results and past events considered by BIDRA. This provides more details of the overall risk picture and approximate assessment results of the real system conditions.

2. Complementarity may reside in the potential of one technique to (partially) validate the other. Despite the possible uncertainty in the definition of variance, the mathematical model of BIDRA is more solid, and it is based on definite events of technical success and failure. On the other hand, the risk barometer uses relatively simple aggregation rules for heterogeneous indicators, organized in a hierarchical structure and weighted on the basis of their relative importance. The definition of this risk model is strongly affected by subjective judgment, and experts should be consulted for most of the risk barometer steps. For this reason, BIDRA results may be compared with risk values from the risk barometer. However, such validation is solely related to the risk barometer's capability to treat technical indicators, because for the sake of consistency, the technique should be deprived of organizational and operational indicators.

4. CONCLUSIONS

Both reactive and proactive approaches aim to provide dynamic assessment of the overall system risk by following partially different paths.

The reactive approach is represented by the BIDRA technique, which relies on solid mathematical modeling and relatively certain inputs. However, it is not exempt from a certain level of subjectivity and does not address potential underlying factors in accidents.

The risk barometer is a proactive technique in which the model structure is relatively adaptable and case specific; however, it is not grounded on rigorous aggregation rules. Its inputs are potentially able to cover a wide umbrella of organizational and operational factors as well as technical issues and past events and allow for deeper insight into accident causes.

Comparison of the two techniques has shown that they do not strictly belong to one of the two classes, but they both possess reactive and proactive traits to some extent. Moreover, despite the fact that they may show little complementarity, their overlapping may be useful for (partial) mutual validation.

REFERENCES

[1] Kalantarnia M, Khan F, Hawboldt K. Dynamic risk assessment using failure assessment and Bayesian theory. Journal of Loss Prevention in the Process Industries 2009;22:600−6.

[2] Kalantarnia M, Khan F, Hawboldt K. Modelling of BP Texas City refinery accident using dynamic risk assessment approach. Process Safety and Environmental Protection 2010;88:191−9.

[3] Hauge S, Okstad E, Paltrinieri N, Edwin N, Vatn J, Bodsberg L. Handbook for monitoring of barrier status and associated risk in the operational phase. SINTEF F27045. Trondheim: Norway Center for Integrated Operations in the Petroleum Industry; 2015.

[4] Paltrinieri N, Scarponi GE, Khan F, Hauge S. Addressing dynamic risk in the petroleum industry by means of innovative analysis solutions. Chemical Engineering Transactions 2014:451−6.

[5] ISO. Risk management—principles and guidelines. ISO 31000:2009. Geneva, Switzerland: International Organization for Standardization; 2009.

[6] Aven T. Practical implications of the new risk perspectives. Reliability Engineering and System Safety 2013;115:136−45.

[7] Aven T, Krohn BS. A new perspective on how to understand, assess and manage risk and the unforeseen. Reliability Engineering & System Safety 2014;121:1−10.

[8] Reuters T. ESG Research Data 2016.

[9] Health and Safety Executive. Developing process safety indicators. HSE guidance series, HSG254. Bootle, United Kingdom: HSE; 2006.

[10] Øien K, Massaiu S, Ranveig KT. Guideline for implementing the REWI method. Trondheim, Norway: SINTEF; 2012.

[11] Schönbeck M, Rausand M, Rouvroye J. Human and organisational factors in the operational phase of safety instrumented systems: a new approach. Safety Science 2010;48:310−8.

[12] Paltrinieri N, Hokstad P. Dynamic risk assessment: development of a basic structure. In: Safety and reliability: methodology and applications − proceedings of the European Safety and Reliability Conference, ESREL 2014; 2015. p. 1385−92.

[13] Saaty TL. Priority setting in complex problems. IEEE Transactions on Engineering Management 1982:140−55. EM-30.

[14] Bana E, Costa CA, De Corte JM, Vansnick JC. On the mathematical foundations of MACBETH. International Series in Operations Research and Management Science 2016:421−63.

[15] Gran BA, Bye R, Nyheim OM, Okstad EH, Seljelid J, Sklet S, et al. Evaluation of the Risk OMT model for maintenance work on major offshore process equipment. Journal of Loss Prevention in the Process Industries 2012;25:582−93.

[16] Øien K. Risk indicators as a tool for risk control. Reliability Engineering and System Safety 2001;74: 129−45.

[17] Øien K. A framework for the establishment of organizational risk indicators. Reliability Engineering and System Safety 2001;74:147−67.

Analysis of Consequences

CHAPTER 9

Dynamic Consequence Analysis through Computational Fluid Dynamics Modeling

G. Landucci[1], M. Pontiggia[2], N. Paltrinieri[3,4], V. Cozzani[5]
[1]University of Pisa, Pisa, Italy; [2]D'Appolonia S.p.A., San Donato Milanese (MI), Italy; [3]Norwegian University of Science and Technology (NTNU), Trondheim, Norway; [4]SINTEF Technology and Society, Trondheim, Norway; [5]University of Bologna, Bologna, Italy

1. INTRODUCTION

Risk analysis involves the estimation of accident consequences using engineering and mathematical techniques [1]. Consequences are aimed at determining the contribution of magnitude to the risk, representing the impact of the expected accidents. Magnitude is related to both vulnerability of territory and severity of scenarios; hence it is often considered a static term associated with the potential hazard in conventional quantitative risk assessment (QRA) frameworks.

Moreover, in dynamic risk assessment (DRA), the magnitude of accidents is not usually involved in the update process, because DRA studies are meant to reassess risk in terms of updating initial failure probabilities of events (causes) and safety barriers as new information is made available during a specific operation [2]. Bayesian [3] and non–Bayesian approaches [4] are thus focused on the dynamic improvement in likelihood of accidents (ie, frequency) rather than focusing on the dynamic variation in accident effects, also accounting for possible mitigation action of safety barriers.

The transient evolution of an accident plays an important role in the estimation of the magnitude, because obstacles, change in meteorological conditions, and site-specific factors connected to the layout of equipment may alter the nature of the impact associated with an accident. Moreover, the intervention of multiple safety layers or barriers may reduce the physical effects associated with accident scenarios.

Computational fluid dynamics (CFD) modeling is a consolidated tool to support industrial project development and was recently adopted in the framework of consequence assessment and safety studies [5]. The advanced features of CFD models, such as handling complex three-dimensional geometries and environments and analyzing reactive or nonreactive flow of compressible or noncompressible fluids, make them a promising tool to support dynamic consequence assessment in the perspective of implementation in DRA studies.

Dynamic Risk Analysis in the Chemical and Petroleum Industry
ISBN 978-0-12-803765-2

2. IMPLEMENTATION OF COMPUTATIONAL FLUID DYNAMICS MODELING INTO DYNAMIC RISK ASSESSMENT

CFD models solve Navier—Stokes equations of fluid flow (conservation of mass, momentum, and scalar quantities) in a three-dimensional space [6]. The problem is reduced to the solution of a system of partial-differential (or integral-differential) equations that need to specify appropriate boundary conditions and discretization methods for their numerical solution. Boundary conditions allow determination of the possible energy and/or mass inlet/outlet and relevant features of the system under analysis (ie, obstacles or other walls that alter the nature of the flow). The discretization method approximates the differential equations in a system of algebraic equations, which can be solved on high-performance computers. In typical commercial codes, the finite volumes approach is adopted. In this approach, the geometrical domain under analysis is first subdivided into a computational grid (ie, the mesh), which allows determining control volumes when the governing equations (ie, conservation of mass, energy, and momentum and the radiative heat transfer equation) are solved.

Therefore, CFD models may capture the interaction among the released hazardous substances and the geometry of the equipment, pipes, and structures, as well as topography and vegetation surrounding an industrial site where risk assessment is carried out. The advantage of CFD is also related to the dynamic nature of the results, which can be used to trace the transient evolution of the physical effects associated with the accident scenario.

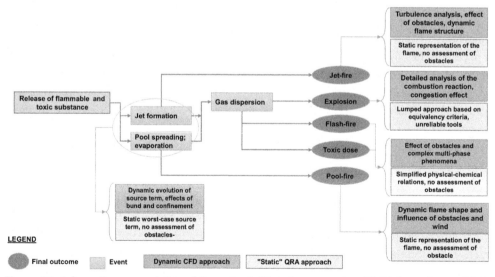

Figure 9.1 Schematic representation of the potentialities of computational fluid dynamics models in capturing dynamic accident scenarios and limitations of lumped parameters models implemented in quantitative risk assessment.

Although consequence assessment may provide detailed and dynamic results through CFD, the evaluation of accident scenarios in typical QRA is based on conservative simplifications and a rather static approach [7]. This is due to the large number of scenarios considered. Disregarding the dynamic nature of the accidental release may potentially lead to overestimation of the release severity. Fig. 9.1 illustrates a scheme in which the application of CFD models is associated with each relevant scenario, also highlighting the limitations of conventional static approaches.

Finally, it is worth noting the possible limitations of CFD codes. Models need to be specifically tuned for the case under analysis, and the related calculations may need high-computational resources. Specific validation against experimental data should be carried out for reliability and accuracy. However, experimental studies are few and require relevant economic resources. The general application process is time consuming and requires skilled users. Owing to these limitations, CFD application in conventional QRA studies is not yet consolidated [8], but the advanced features of the results make CFD an attractive tool for supporting DRA studies.

3. COMPUTATIONAL FLUID DYNAMIC MODELS FOR SPECIFIC ACCIDENTAL SCENARIOS

3.1 Advanced Source Terms and Outflow Models

Release source terms are used to quantify the flow rate at which accidental releases of dangerous substances may occur and/or to estimate release quantity and duration. In typical QRA studies, source terms and outflow models are simplified expressions, often providing rough and static worst-case estimations. Advanced CFD tools may predict the dynamic development of the release and complicating phenomena resulting from obstacles or change in wind conditions, thus supporting dynamic consequence assessment.

3.1.1 Computational Fluid Dynamic Modeling of Jet Releases

The assessment of jet releases is a challenging task because of the likelihood of high turbulence and complex premixing phenomena. Moreover, jet releases contacting the ground and obstacles may be altered in structure and effects. CFD studies may achieve detailed prediction of jet structure in terms of pressure, temperature, and composition profiles able to support advanced dispersion studies in the presence of obstacles or wall barriers. In the past decade, safety studies in the literature have focused on high-momentum jet releases of mainly methane and hydrogen [9]. More recently, in the perspective of the development of carbon capture and sequestration technologies, CFD models were adopted to predict the complex phenomena associated with accidental carbon dioxide release from pipelines or transport tankers. Gant et al. [10] investigated the formation of a carbon dioxide jet from a pressurized vessel, accounting for the formation of sublimating particles as well as for the contribution of moisture in the entrainment of water droplets in the jet. The sound modeling of carbon dioxide jet behavior is crucial for supporting (1) the tracing of the three-phase flow and (2) the following toxic dispersion studies (such as the ones discussed in Section 3.2).

3.1.2 Computational Fluid Dynamic Modeling of Pool Spreading and Evaporation

Liquid spillage due to failure of equipment and pipelines may lead to the spreading of a pool and its vaporization. This stage of an accident scenario provides important information for the subsequent analysis of atmospheric dispersion. Additional criticality is associated with pool spreading, especially for cryogenic liquids. In fact, the effects of terrain, release characteristics, and obstacles are critical in the modification of heat and material balance.

Relevant studies concerning the application of CFD models to cryogenic pool spreading and evaporation were concentrated on liquefied natural gas (LNG) and liquefied hydrogen (LH_2). Gavelli et al. [11] addressed the prediction of LNG behavior in an impoundment through CFD modeling implemented on ANSYS Fluent software. A critical issue in the modeling was the characterization of the LNG spill, because the evaluation of spill velocity is crucial for the determination of turbulent kinetic energy associated with the pool spread and its propensity to entrain air. GexCon As (Bergen, Norway) showed the capabilities of the flame acceleration simulator CFD code in predicting LH_2 release evolution to support dispersion and explosion studies through the introduction of dedicated submodels for determining thermal and kinematic equilibrium among liquid and vapor phases [12].

3.2 Dispersion Studies

Risk assessment of accidental releases of hazardous gases, both flammable and toxic, is largely performed using integral dispersion models, the most widely used being DEGADIS, SLAB, ALOHA, and Unified Dispersion Model (UDM) [8]. Integral models are lumped-parameter models that provide reliable results only in open field conditions, that is, when almost no obstacles are present in the cloud region. Moreover, the source term providing the dispersed flow rate is usually assumed to be constant [1,7]. When complex situations are analyzed, the adoption of CFD is needed to perform a dynamic three-dimensional simulation of the involved geometry. Typical examples are the analysis of congested industrial layouts with neighboring equipment, pipes, or obstacles [13–15] or an urban environment (buildings and street canyons) [16–19]. The following sections address representative studies of complex environment CFD dispersion, considering both flammable and toxic gases. Moreover, an approach for the selection of suitable modeling strategies is discussed.

3.2.1 Computational Fluid Dynamic Modeling of Flammable Gas Dispersion

The aim of dispersion studies of flammable materials is the evaluation of the potential flammable cloud extension and features, that is, the zone in which the concentration of the released gas is within the flammable limits [7]. As mentioned in Section 2, this is crucial for the assessment of flash fire and vapor cloud explosion scenarios.

Several examples are available in the literature, mainly focusing on natural gas dispersion following accidental LNG spills [14,15]. This is a complex phenomenon requiring a dedicated preliminary assessment because it involves the formation of a liquid pool that spreads and evaporates. The natural gas evaporated from the pool is at extremely low temperatures (about 112–115 K), implying a density higher than the air density (heavy gas dispersion). The gas is stratified on the ground and influenced by obstacles close to the release point.

The presence of obstacles may lead to scenario mitigation because of the turbulence increment in the gas cloud, which increases its dilution with air, and the warming due to the heat exchange with the atmosphere. This is considered a key strategy for reducing the hazards associated with heavy gas dispersion, as pointed out by Busini and Rota [20]. Relevant CFD studies of heavy gas dispersion were also undertaken for the consequence assessment of the LPG flash fire that occurred in Viareggio, Italy, in 2009 [18,19]. More details on this issue are extensively covered in Chapter 10.

The release of light buoyant gas in the presence of high-momentum jet releases was undertaken in similar studies carried out by Wilkening and Baraldi [9], which simulated the dispersion of gaseous H_2 and CH_4 from the pipe rupture in a refueling station located in a residential area.

3.2.2 Computational Fluid Dynamic Modeling of Toxic Gas Dispersion

Several CFD dispersion studies were carried out to capture the effect of toxic substances dispersed over congested and complex areas, especially in the framework of transportation risk assessment studies. In this case, toxic concentrations such as emergency response planning guidelines, immediately dangerous to life and health concentration, and

Table 9.1 Summary of Relevant Computational Fluid Dynamics Dispersion Studies Carried Out for the Assessment of Toxic Effects

ID	References	Substance	Release Source	Type of Environment
A	[13]	Hydrogen chloride	Accidental leak from process pipe	Industrial area (chemical plant close to a nuclear plant)
B	[16]	Ammonia–water solution (10%, w/w)	Catastrophic rupture of road tanker	Urban area (Lecco, Italy)
C	[17]	Chlorine	Punctured railroad car	Industrial park (Festus, US); urban area (Chicago, US)

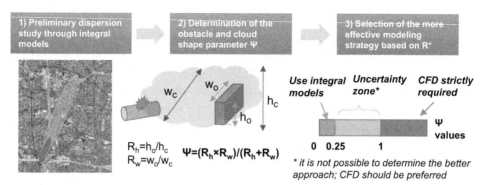

Figure 9.2 Methodology for the selection of the more effective modeling strategy for dispersion studies in the presence of obstacles. *(Adapted from Derudi M, Bovolenta D, Busini V, Rota R. Heavy gas dispersion in presence of large obstacles: selection of modeling tools. Industrial & Engineering Chemistry Research 2014;53:9303−10.)*

lethal concentration, 50% (see Ref. [7] for more details) and/or toxic doses are traced to assess the potential damages associated to toxic cloud dispersions. Table 9.1 summarizes relevant examples carried out with different substances and in different types of zones (ie, industrial and urban).

3.2.3 Guidelines for the Selection of the Most Suitable Modeling Strategy

The collected CFD studies are relevant examples demonstrating the potential of CFD models in assessing complex dispersion scenarios. Nevertheless, the usage of a CFD tool must be restricted to cases significantly influenced by geometry, because it could lead to a waste of computational resources in the case of relatively simple geometries. For this reason, a methodology based on the comparison among CFD and integral model simulations was established to drive the selection of the proper modeling strategy. More details are reported by Derudi et al. [21]. The methodology, which is based on the preliminary simulation though integral models and on the characterization of defects in the area of interest, is summarized in Fig. 9.2.

3.3 Fire Studies

Fire dynamic simulation is of utmost importance for assessing the severity of fire events following the accidental ignition of flammable substances. CFD models may significantly contribute to the assessment of heat radiation effects as well as to the analysis of fire-driven fluid flow. Moreover, convective effects associated with direct flame engulfment and development of smoke and toxic gases may be traced.

3.3.1 Computations Fluid Dynamic Modeling of Dynamic Fire Scenarios

Conventional QRA studies are carried out assuming a static geometry of the flame and, consequently, steady state effects. In fact, pool fires and jet fires are simulated without accounting for complicating phenomena, which may alter the shape of the flame and the heat radiation effects on potential targets.

Pool fire modeling through CFD has been extensively carried out since the 1990s, determining the potentialities of distributed parameter codes in capturing the effects of bunds, wind profiles, and confinement in the determination of flame structure and associated effects [22]. More recently, Sun and Guo [23] provided a dynamic LNG pool fire simulation comparing the effect with and without mitigation through high-expansion foam at different burning times. CFD studies were adopted to predict the effectiveness of the foam and to provide indication of optimal foam-delivering conditions.

As a result of higher turbulence, jet fire modeling is a more crucial task and was improved in recent years. Wang et al. [24] adopted FireFOAM to study the radiation characteristics of hydrogen and hydrogen/methane jet fires, capturing the fluctuations in flame length and radiant fraction.

Jang et al. [25] simulated a hydrogen jet fire from an accidental leak, determining the dynamic evolution of the flame temperature and shape into a complex three-dimensional layout. A real-scale pipe rack was reproduced, determining the flame impact zone as well as the heat radiation profiles.

3.3.2 Assessment of Domino Effect Triggered by Fire

Fires may affect process and storage equipment, causing severe damage and potential accident escalation owing to the domino effect [26].

CFD models may be set up to simulate pressurized vessels exposed to accidental fires, determining the transient behavior of the stored fluid during heat-up. CFD allows for the prediction of velocity and temperature profiles, obtaining the pressurization rate in the

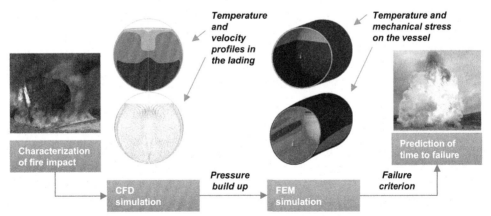

Figure 9.3 Advanced modeling approach to support the assessment of a domino effect triggered by fire.

vessel and providing key indications for the evaluation of the vessel resistance as well as for the design of heat-resistant coating for fireproofing. The results can then be implemented in finite elements modeling for the assessment of the mechanical response of the structure/equipment affected by the fire, thus determining the time to failure (TTF). TTF is a crucial element for the analysis of domino effect scenarios and for emergency management. Fig. 9.3 summarizes the modeling strategy adopted by Landucci et al. for the integrated assessment of the thermal [27] and mechanical [28] response of pressurized vessels exposed to fires to estimate TTF and to support the advanced assessment of the domino effect triggered by fires.

3.4 Explosion Studies

In recent years, CFD codes became a routine tool for the simulation of the consequences of vapor cloud explosions or partially confined explosions, as in the case of vented equipment or offshore rigs, characterized by a high level of confinement and congestion [29]. Several engineering analyses of large-scale industrial explosions were carried out through the Reynolds-averaged Navier—Stokes (RANS) approach. However, the RANS for the reproduction of large-scale, compressible, reactive systems is questionable, and very large uncertainties are intrinsically produced as the result of several conservative assumptions about the complex combustion phenomena involved in explosions. Recently, the use of large eddy simulation has been proposed. This technique is still under development and is limited by the available computational power [30].

With respect to conventional lumped parameter models adopted for the analysis of explosions, CFD may provide the dynamic development of the pressure wave and of the flame zone, capturing complex phenomena such as the deflagration to detonation transition, which is critical in extremely congested environments [31].

The same type of approach shown in Section 3.3.2 for the domino effect triggered by fires may be extended to the analysis of equipment and pipes affected by a shock wave. In this case, CFD simulation is preliminarily carried out for the assessment of the blast load on the equipment or asset under analysis. More details on the combined CFD—finite element method analysis of explosion effects are reported elsewhere [32].

4. CONCLUSIONS

This chapter presented the application of CFD models for the evaluation of the impact of transient accidental scenarios, in particular considering dispersion, fires, and explosions. Relevant studies and examples highlighted the potentialities in the implementation of DRA studies, because CFD models can capture the presence of obstacles and barriers and reduce the impact of accidents, as well as considering the modification in weather conditions (ie, wind speed and direction, humidity, etc.).

Nevertheless, the complexity of input information and data together with computer time and man-hours required to build the numerical domain pose important limitations for extensive CFD application, which should be limited to more critical cases. Hence, examples driving the choice between lumped models and advanced CFD simulations were provided.

To provide more details on the implementation of CFD models in a dynamic consequence assessment framework, Chapter 10 outlines a tutorial in which a dynamic dispersion study is carried out through the use of a CFD model.

REFERENCES

[1] Crowl DA, Louvar JF. Chemical process safety — fundamentals with applications. 2nd ed. New Jersey: Prentice Hall PTR; 2002.

[2] Abimbola M, Khan F, Khakzad N. Dynamic safety risk analysis of offshore drilling. Journal of Loss Prevention in the Process Industries 2014;30:74—85. http://dx.doi.org/10.1016/j.jlp.2014.05.002.

[3] Kalantarnia M, Khan F, Hawboldt K. Dynamic risk assessment using failure assessment and Bayesian theory. Journal of Loss Prevention in the Process Industries 2009;22:600—6. http://dx.doi.org/10.1016/j.jlp.2009.04.006.

[4] Khakzad N, Khan F, Amyotte P. Dynamic risk analysis using bow-tie approach. Reliability Engineering & System Safety 2012;104:36—44.

[5] Schmidt J. Process and plant safety: applying computational fluid dynamics. Berlin, Germany: Wiley-VCH Verlag; 2012.

[6] Ferziger JH, Peric M. Computational methods for fluid dynamics. Berlin, Germany: Springer-Verlag; 2002.

[7] Lees FP. Loss prevention in the process industries. 2nd ed. Oxford: Butterworth - Heinemann; 1996.

[8] Antonioni G, Burkhart S, Burman J, Dejoan A, Fusco A, Gaasbeek R, et al. Comparison of CFD and operational dispersion models in an urban-like environment. Atmospheric Environment 2012;47:365—72.

[9] Wilkening H, Baraldi D. CFD modelling of accidental hydrogen release from pipelines. International Journal of Hydrogen Energy 2007;32:2206—15.

[10] Gant SEE, Narasimhamurthy VDD, Skjold T, Jamois D, Proust C. Evaluation of multi-phase atmospheric dispersion models for application to Carbon Capture and Storage. Journal of Loss Prevention in the Process Industries 2014;32:286—98.

[11] Gavelli F, Bullister E, Kytomaa H. Application of CFD (Fluent) to LNG spills into geometrically complex environments. Journal of Hazardous Materials 2008;159:158—68.

[12] Ichard M, Hansen OR, Middha P, Willoughby D. CFD computations of liquid hydrogen releases. International Journal of Hydrogen Energy 2012;37:17380—9.

[13] Sully A, Heitsch M, Baraldi D, Wilkening H. Numerical simulations of hydrogen and hydrogen chloride releases in a nuclear hydrogen production facility. International Journal of Hydrogen Energy 2011;36:1083—93.

[14] Luketa-Hanlin A, Koopman RP, Ermak DL. On the application of computational fluid dynamics codes for liquefied natural gas dispersion. Journal of Hazardous Materials 2007;140:504—17.

[15] Sun B, Utikar RP, Pareek VK, Guo K. Computational fluid dynamics analysis of liquefied natural gas dispersion for risk assessment strategies. Journal of Loss Prevention in the Process Industries 2013;26:117—28.

[16] Pontiggia M, Derudi M, Alba M, Scaioni M, Rota R. Hazardous gas releases in urban areas: assessment of consequences through CFD modelling. Journal of Hazardous Materials 2010;176:589—96.

[17] Hanna SR, Hansen OR, Ichard M, Strimaitis D. CFD model simulation of dispersion from chlorine railcar releases in industrial and urban areas. Atmospheric Environment 2009;43:262—70.

[18] Pontiggia M, Landucci G, Busini V, Derudi M, Alba M, Scaioni M, et al. CFD model simulation of LPG dispersion in urban areas. Atmospheric Environment 2011;45:3913–23.

[19] Landucci G, Tugnoli A, Busini V, Derudi M, Rota R, Cozzani V. The Viareggio LPG accident: lessons learnt. Journal of Loss Prevention in the Process Industries 2011;24:466–76.

[20] Busini V, Rota R. Influence of the shape of mitigation barriers on heavy gas dispersion. Journal of Loss Prevention in the Process Industries 2014;29:13–21.

[21] Derudi M, Bovolenta D, Busini V, Rota R. Heavy gas dispersion in presence of large obstacles: selection of modeling tools. Industrial & Engineering Chemistry Research 2014;53:9303–10.

[22] Sinai YL, Owens MP. Validation of CFD modelling of unconfined pool fires with cross-wind: flame geometry. Fire Safety Journal 1995;24:1–34.

[23] Sun B, Guo K. LNG accident dynamic simulation: application for hazardous consequence reduction. Journal of Loss Prevention in the Process Industries 2013;26:1246–56.

[24] Wang CJ, Wen JX, Chen ZB, Dembele S. Predicting radiative characteristics of hydrogen and hydrogen/methane jet fires using FireFOAM. International Journal of Hydrogen Energy 2014;39: 20560–9.

[25] Jang CB, Choi S-W, Baek J-B. CFD modeling and fire damage analysis of jet fire on hydrogen pipeline in a pipe rack structure. International Journal of Hydrogen Energy 2015;40:15760–72.

[26] Landucci G, Cozzani V, Birk M. Heat radiation effects. In: Domino Eff. Process Ind. Model. Prev. Manag. Amsterdam, The Netherlands: Elsevier; 2013. p. 70–115.

[27] D'Aulisa A, Tugnoli A, Cozzani V, Landucci G, Birk A. CFD modeling of LPG vessels under fire exposure conditions. AIChE Journal 2014;60:4292–305.

[28] Landucci G, Molag M, Cozzani V. Modeling the performance of coated LPG tanks engulfed in fires. Journal of Hazardous Materials 2009;172:447–56.

[29] Lea CJ, Ledin HS. A review of the state-of-the-art in gas explosion modelling (Report HSL/2002/ 02). Buxton, UK: Health and Safety Laboratory, Fire and Explosion Group; 2002.

[30] Di Sarli V, Di Benedetto A, Russo G. Sub-grid scale combustion models for large eddy simulation of unsteady premixed flame propagation around obstacles. Journal of Hazardous Materials 2010;180: 71–8.

[31] Hansen OR, Johnson DM. Improved far-field blast predictions from fast deflagrations, DDTs and detonations of vapour clouds using FLACS CFD. Journal of Loss Prevention in the Process Industries 2015;35:293–306.

[32] Landucci G, Salzano E, Taveau J, Spadoni G. Domino effects in the process industries. Elsevier; 2013.

CHAPTER 10

Computational Fluid Dynamics Modeling: Tutorial and Examples

G. Landucci[1], M. Pontiggia[2], N. Paltrinieri[3,4], V. Cozzani[5]

[1]University of Pisa, Pisa, Italy; [2]D'Appolonia S.p.A., San Donato Milanese (MI), Italy; [3]Norwegian University of Science and Technology (NTNU), Trondheim, Norway; [4]SINTEF Technology and Society, Trondheim, Norway; [5]University of Bologna, Bologna, Italy

1. INTRODUCTION

In this chapter, the potential of computational fluid dynamics (CFD) simulations in supporting dynamic risk assessment is presented. The aim is to demonstrate the advanced features of CFD tools, allowing for more comprehensive analysis of accident scenarios with respect to conventional models based on integral approaches.

This chapter is devoted to the assessment of release and dispersion of liquefied petroleum gas (LPG) in an urban context through the analysis of a specific case history: The accident in Viareggio, Italy, in 2009 is used as a reference case study [1]. The accident followed a freight train derailment in which an LPG tank car was punctured, releasing its entire content. The ignition led to a severe flash fire, with consequent extended property damages and 32 fatalities.

The methodology adopted for setting up the CFD model is shown to highlight the complexity of the tool as well as the level of detail needed to carry out a real-scale consequence assessment study. The application of conventional integral models, in particular the unified dispersion model (UDM) [2], is presented as well. The results of the calculation are compared with the reported damages, presenting indications for the selection of the proper modeling strategy in the perspective of dynamic risk assessment.

2. METHODOLOGY TUTORIAL

Fig. 10.1 shows the flowchart of the approach and generic requirements needed to perform the CFD study. The figure also shows the alternative path needed for setting up the same type of simulation by adopting integral tools, which reduces the simulation to only two steps (ie, steps 1 and 5 in Fig. 10.1).

The preliminary step (step 0) of a CFD study consists of the determination of the scenario to be analyzed. This is a crucial step because setting up CFD requires relevant resources, in terms of both man-hours and computational resources. Chapter 9 discusses the selection criteria for a proper modeling strategy to limit the number of cases to be assessed with CFD tools. Hence, the preliminary phase is out of the scope of this chapter.

Dynamic Risk Analysis in the Chemical and Petroleum Industry
ISBN 978-0-12-803765-2

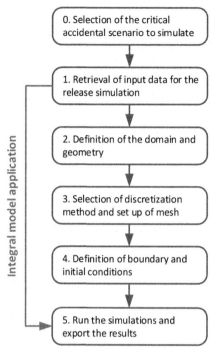

Figure 10.1 Flowchart of the methodology.

This tutorial focuses on the application of CFD models for the analysis of a scenario: flammable gas dispersion in complex environment, such as urban areas, by performing a three-dimensional study. The aim of dispersion studies for flammable materials is the evaluation of the extension and features of the flammable cloud, that is, the zone in which the concentration of the released gas is between the flammable limits (lower flammability limit [LFL] and upper flammability limit [UFL]) [3].

The first step (step 1) is to retrieve input information to (1) characterize the surrounding environment and (2) determine the source term of the released substance. The surrounding environment needs to be characterized by determining the meteorological conditions during the dispersion scenario and the features of the territory in which the dispersion takes place. Wind and atmospheric stability play an important role in the dispersion study (see Lees [3] for more details) and may significantly change, affecting the direction and shape of the flammable cloud. For evaluation of the released substance source term, information about the release size (eg, the equivalent release diameter) and operative data prior to the loss of containment (eg, pressure, temperature, composition, and inventory of the hazardous substance) from a tank or other detention system are needed [3,4]. Based on this information, CFD tools may simulate the outflow of the substance, or the source term can be directly set up by the user as an "input mass" condition [5]. In the latter case, integral models for source terms [4] may also be used to limit the computational resources for the CFD simulation, as explained in Section 3.2.1.

As shown in Fig. 10.1, the information in step 1 constitutes the basis for performing simulation through integral modeling, thus obtaining the results in terms of iso-concentration contours. Instead, CFD models need several additional steps, which are described in detail for the case under analysis.

Step 2 is aimed at defining the domain in which the simulation needs to be carried out, obtaining a three-dimensional spatial representation of the geometrical environment. ANSYS ICEM (ANSYS, Inc., Canonsburg, PA) [5] or other generic computer-assisted design tools may be adopted for defining and drawing the geometry. Step 2 supports the definition of the computational grid (step 3), which is needed in commonly applied software for CFD modeling (such as ANSYS Fluent (ANSYS, Inc., Canonsburg, PA) [5]) adopting the finite volume approach. The computational grid (or mesh) is obtained from the subdivision of the three-dimensional space into control volumes, in which the governing equations (conservation of mass, energy and momentum and radiative heat transfer equation) are solved. In this step, solution and discretization methods are also selected, depending on the type of problem, to achieve the optimal and stable solution convergence. More details on this topic are extensively reported elsewhere [5–8].

Once the numerical set-up of the problem is carried out, boundary conditions need to be specified. Boundaries direct motion of flow, specifying solid (eg, ground, obstacles, buildings, and vegetation) and fluid zones (in the case of a dispersion study, the portion of the domain occupied by atmospheric air and the gas-flammable or toxic substance). Typical boundary conditions for dispersion studies are summarized in Table 10.1.

Finally, the simulation can be run to obtain the final results (step 5). As mentioned earlier, the key outcome of a dispersion study of flammable gases is the obtainment of a dynamic concentration profile. With respect to integral models, CFD allows advanced types of profiles, such as streamlines or velocity vectors distribution. This allows the obtaining of information on the influence of obstacles on fluid motion.

3. APPLICATION OF THE APPROACH

3.1 Description of the Case Study

In this section, the potentialities for CFD in analyzing the dynamic evolution of a large-scale flammable cloud in a congested urban area are shown through the analysis of a case study. In particular, the consequence analysis of the release event following the Viareggio train accident is presented.

On June 29, 2009, a freight train was carrying 14 LPG tank cars through the densely populated area of Viareggio at a speed of 90 km/h. A tank car derailed, overturned, and was punctured, releasing its entire content. LPG vaporized and formed a cloud that spread over the railway area and extended over a residential zone near the railway. A flash fire followed the cloud ignition, causing extended damages [9]. In some locations, LPG vapors penetrated into enclosed zones, such as houses and an underground garage, leading to a confined explosion following ignition.

Table 10.1 Examples of Boundary Conditions Adopted in Gas Dispersion Studies Carried Out by Computational Fluid Dynamics Models

Type of Element	Schematization	Type of Boundary Condition
Inlet of the substance in the domain		Because the gas is characterized by compressible flow, mass or pressure inlet can be typically adopted [5]. Source term may be calculated with integral models [4] and imposed by the user to limit computational efforts. Temperature and turbulence of the inlet mass are assigned as well.
Outlet from the domain		Pressure outlet is typically adopted, defining the static/gauge pressure at the outlet boundary. This is interpreted as the static pressure of the environment into which the flow exhausts.
Surrounding air and wind inlet		As a result of low speed (subsonic conditions), incompressible flow may be assumed, imposing a velocity profile, temperature, and turbulence values for the inlet wind flux.
Ground and obstacles		No-slip boundary condition is commonly applied considering a solid "wall" for the ground, obstacles, buildings, etc. This is used to bound fluid and solid regions. Roughness and temperature are also specified.

Severe flame exposure   ***Moderate flame exposure***   ***Radiation exposure***

Figure 10.2 Example of classification of damaged vehicles and buildings with increasing level of damage.

A detailed analysis of the consequences of the fire has also been carried out, taking into account a systematic classification of damages. The following types of target have been considered: structures/houses, vehicles, and vegetation. Three types of damage, with increasing severity, have been identified [1]: (1) radiation damage due to distant radiation; (2) moderate flame damage due to flame engulfment with significant effects on grass/hedges and poor effects on buildings; and (3) severe flame damage (grass/vegetation set on fire, buildings or cars set on fire, rupture of windows due to strong thermal dilatation). For the sake of brevity, an example of the increasing level of damages is shown for vehicles and buildings in Fig. 10.2.

This series of events having extremely severe consequences is a typical example of the hazard posed by accident scenarios in LPG transportation by road and rail [10,11]. Dynamically modeling the consequences of such accidental scenarios is the only way to foresee the impact area to design appropriate mitigation measures and to prepare specific emergency plans based on the expected impact area of such accidents [12,13].

To reproduce the effects of the accident, a CFD model was applied. The model performance was analyzed in detail by a critical comparison of CFD simulation results with the damage map obtained by analyzing the effects of fire on buildings, vehicles, and vegetation [1]. More details about model development and set-up are reported elsewhere [1,9,14]. Here the relevant aspects are summarized to highlight the possible advanced features associated with the implementation of a CFD model to analyze the dynamic evolution of accident scenarios. Moreover, the set-up of conventional integral models [2] is shown, and the results of the two modeling approaches are compared.

3.2 Simulation Settings

3.2.1 Computational Fluid Dynamics Model

The geometry of the area of interest was imported into the CFD code from a topographical database available for Viareggio [15]. The database allowed fast and easy geometry generation: The perimeter at the eaves' height and the position were available for each building, and buildings were simulated extruding top area to the ground, leading to simplified representation of the urban terrain as a combination of parallelepiped structures. Edges smaller than 0.2 m, that is, the smaller dimension documented in the topographical database, were smoothed to remove relatively small details that would require high grid refinement but would play a minor role in gas dispersion. Further details on the geometry import procedure can be found elsewhere [16].

The domain area is represented in Fig. 10.3A, featuring a total extension of $350 \times 500 \times 45$ m. Fig. 10.3 displays the schematization used to represent the actual geometry in the CFD code. Fig. 10.3B shows some details added to the imported geometry, such as containment walls present along the railway as well as the footbridge. Wall thickness was assumed constant and equal to 0.2 m, and the slope of the stair flights was set equal to 45 degree to reduce mesh skewness. Trees were added in the most densely packed zone (see an example in Fig. 10.3B), whereas isolated trees were not considered owing to their low influence on the dispersion of the gas. Average tree dimensions were obtained through direct onsite measurement and were used in the geometry generation. Each single tree was modeled as a solid body, therefore being not permeable to the gas.

Tank cars (apart from the punctured one) and train engine were modeled as boxes 16 m long, 3 m wide, and 3 m high. The overview of the tank car's position is reported in Fig. 10.3A. The punctured tank car was reproduced with a higher level of detail. A cylindrical tank and wheel axle encumbrance were represented. Roadbed was also reproduced, imposing a standard slope and width. Details of tank car geometries are reported in Fig. 10.3B.

Buildings were considered solid; therefore, no gas penetration was simulated. The garage in which a partially confined explosion occurred was represented as a cavity in the ground. Access pads were also reproduced to investigate the behavior of the dense cloud in correspondence with the slope and to check the capability of the CFD code to forecast a gas infiltration large enough to justify the gas explosion (see Fig. 10.3C).

From the punctured tank, a flashing liquid was assumed to enter the atmosphere, leading to the formation of a gas jet (from the flash fraction) and of an evaporating pool (from the liquid rainout). Gas inlet surface (into the integration domain) was obtained on the lateral surface of the cylindrical tank, at its axial end, facing downward and impinging on the ground, in agreement with reported tank damage and with the tank position after the accident. PHAST software (DET NORSKE VERITAS (DNV), London, UK) [17] was adopted for both jet and pool simulation, assuming saturated liquid at atmospheric temperature inside the tank car. The CFD simulation boundary conditions were in this case preelaborated through integral modeling to optimize the computational resources.

A triangular unstructured grid was imposed over the buildings, with a characteristic length of half the smaller edge up to a maximum of 1 m. A 0.05-m triangular grid was imposed on the gas inlet surface, whereas a 0.2-m grid was used at ground level near the source to describe pool evaporation. A size function was set with a growth factor of 1.2 and a size limit of 20 m, leading to a tetrahedral unstructured grid of about 8 million elements (see Fig. 10.3D and E for some mesh details). Each simulation required about 40 h on a 16-parallel-processes cluster.

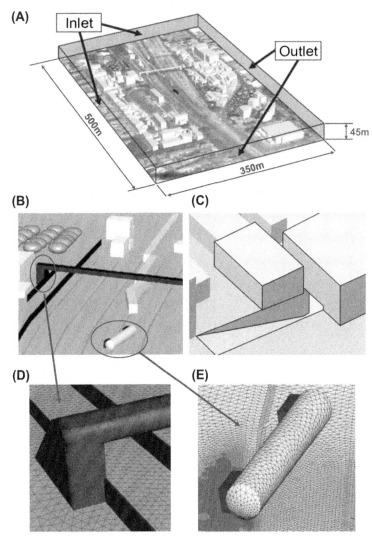

Figure 10.3 Geometry and mesh adopted for the simulation of Viareggio liquefied petroleum gas dispersion. (A) Overview of the domain and boundary definition; (B) schematization of footbridge, walls, trees, and train added to the imported geometry; (C) details of buildings and of an underground garage; (D) mesh definition details for the footbridge; and (E) mesh definition details for the derailed tank car.

3.2.2 Integral Models

In the analysis of potential consequences of hazardous material releases carried out for quantitative risk assessment, simplified models are used owing to the wide areas to analyze and the uncertainty affecting the release position and features [18]. Thus, in general, source-point concentrated parameter dispersion models are used to model gas dispersions.

Dispersion simulations were carried out using UDM as implemented in the PHAST software. More details on implementation and UDM description are reported elsewhere [17]. Two main alternative simulation hypotheses were considered in the present study: (1) simulation of a continuous horizontal LPG release considering an average surface roughness typical of building area (= 1 m), and (2) puff release of 25,000 kg of LPG (initial condition at t_D is stoichiometric mixture with air), with low wind speed (2 m/s in stability class F, blowing from west to northwest).

3.3 Results and Discussion

Fig. 10.4 shows the results of the CFD simulation at 300 s. In Fig. 10.4A, model predictions have been superimposed over the maps of the observed damages (iso-concentration footprint), classified according to the rules defined in Section 3.1. The figure shows that almost all the "severe" and "moderate" damage points are within (or very close to) the

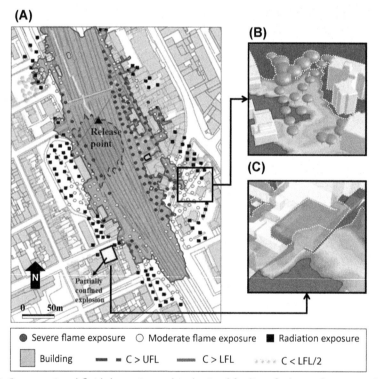

Figure 10.4 Computational fluid dynamic results obtained for liquefied petroleum gas dispersion in urban area. (A) Predicted extension of flammability limits 300 s after the release compared to observed damages; (B) details of the dispersion study on the east side of the station; (C) details of the dispersion study on the west side of the station showing the flammable cloud descending into the garage where a confined explosion occurred. C, concentration of liquefied petroleum gas; LFL, lower flammability limit; UFL, upper flammability limit.

predicted LFL boundary, therefore validating the predictive capabilities of the CFD model used.

Few severe and moderate damage points lay outside the LFL boundary. Nevertheless, those are included in the zone where concentrations are above the value of LFL/2. This boundary, which is only a few meters wider than the LFL boundary, is proposed by several technical standards to conservatively identify the maximum extension of the region affected by flash fire damage. As an example, Fig. 10.4B shows the concentration profiles at ground level in a garden on the east side of the railway (see Fig. 10.4A for the location in the area), where moderate flame exposure was identified. As shown in the figure, although the predicted concentrations are lower than the LFL boundary, almost all the garden is adopted by predicted concentration values between LFL and LFL/2, which fully envelope all the severe and moderate damage points.

Similarly, Fig. 10.4C shows the concentration profiles close to the underground garage in which a partially confined explosion was reported (see Fig. 10.4A for the location in the area). In the figure, the walls of the buildings over the garage were rendered partially transparent to allow the visualization of the concentration field inside the garage. The predicted concentration values in the garage are larger than LFL/2 limits, and the LFL boundary is only a few meters from the observed damages. If compared with the cloud dimensions, which are on the order of a few hundred meters, there is good agreement between the reported damages and the CFD model simulation.

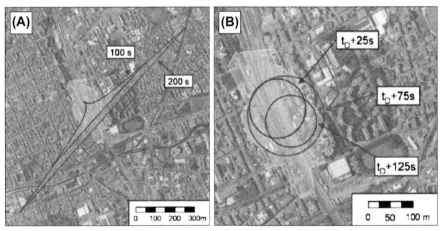

Figure 10.5 Examples of results in the simulation for the Viareggio dispersion by alternative simulation hypothesis. (A) Lower flammability limit per two contours for continuous horizontal release, surface roughness typical of building area ($z_0 = 1$ m); (B) lower flammability limit contours at three simulation times for puff release of 25,000 kg of liquefied petroleum gas (initial condition at t_D is stoichiometric mixture with air). *Yellow marks* (white in print versions) the area actually intersected by damage in the accident. Wind direction was west to northwest. *(Adapted from Landucci G, Tugnoli A, Busini V, Derudi M, Rota R, Cozzani V. The viareggio LPG accident: lessons learnt. Journal of Loss Prevention in the Process Industry 2011;24:466–476. doi:10.1016/j.jlp.2011.04.001.)*

In the case of a simulation carried out with integral models, the results are rather different from the actual records. Fig. 10.5 reports some results obtained for different simulations carried out using UDM as implemented in the PHAST software [17]. As shown in the figure, different assumptions about release conditions and atmospheric dispersion (introduced to allow the application of these simplified models) lead to quite different results. All the alternative approaches lead to conservative results. The results can also be very conservative, leading to very large impact areas with a consequent waste of resources when emergency plans have to be prepared.

Moreover, it is worth mentioning that CFD allows identifying and analyzing a scenario that cannot be captured by conventional static approaches. In particular, as shown in Fig. 10.4C, because flammable concentrations are predicted in the underground garage, the possibility of confined explosion introduces a novel scenario besides the flash fire, thus demonstrating the capabilities of CFD in supporting dynamic hazard and risk assessment.

4. CONCLUSIONS

In this tutorial, the potential of CFD models in simulating large-scale dispersion scenarios is analyzed in the perspective of dynamic risk assessment. In particular, the consequences of the Viareggio accident involving a flash fire due to LPG release and vaporization following a train derailment have been simulated in detail.

The good agreement found between the actual observed damages and the predictions obtained using the CFD approach strongly supports the reliability of the CFD approach for simulating heavy gas dispersion in geometrically complex environments. Moreover, the results shown with the static integral model approach appear extremely conservative, leading to an overestimation of consequences in the far field.

The advanced features of CFD models may foresee the evolution of accidental scenarios that may evolve into additional accidents, such as the confined explosion in the garage captured by the dispersion study, demonstrating the importance of CFD studies in the framework of dynamic consequence and risk assessment.

REFERENCES

[1] Landucci G, Tugnoli A, Busini V, Derudi M, Rota R, Cozzani V. The viareggio LPG accident: lessons learnt. Journal of Loss Prevention in the Process Industry 2011;24:466−76. http://dx.doi.org/10.1016/j.jlp.2011.04.001.

[2] Antonioni G, Burkhart S, Burman J, Dejoan A, Fusco A, Gaasbeek R, et al. Comparison of CFD and operational dispersion models in an urban-like environment. Atmospheric Environment 2012;47:365−72.

[3] Lees FP. Loss prevention in the process industries. 2nd ed. Oxford: Butterworth-Heinemann; 1996.

[4] van den Bosch CJH, Weterings RAPM. Methods for the calculation of physical effects (Yellow Book). third edition. The Hague (NL): Committee for the Prevention of Disasters; 2005.

[5] ANSYS Inc.. ANSYS® FLUENT® 14.5 theory guide. Cecil Township, PA: ANSYS Inc.; 2012.

[6] Schmidt J. Process and plant safety: applying computational fluid dynamics. Berlin, Germany: Wiley-VCH Verlag; 2012.

[7] Ferziger JH, Peric M. Computational methods for fluid dynamics. Berlin, Germany: Springer-Verlag; 2002.

[8] Lomax H, Pulliam T, Zingg D, Kowalewski T. Fundamentals of computational fluid dynamics. Applied Mechanics Review 2002;55:B61. http://dx.doi.org/10.1115/1.1483340.

[9] Pontiggia M, Landucci G, Busini V, Derudi M, Alba M, Scaioni M, et al. CFD model simulation of LPG dispersion in urban areas. Atmospheric Environment 2011;45:3913−23. http://dx.doi.org/10.1016/j.atmosenv.2011.04.071.

[10] Paltrinieri N, Landucci G, Molag M, Bonvicini S, Spadoni G, Cozzani V. Risk reduction in road and rail LPG transportation by passive fire protection. Journal of Hazardous Material 2009;167:332−44. http://dx.doi.org/10.1016/j.jhazmat.2008.12.122.

[11] Bubbico R, Ferrari C, Mazzarotta B. Risk analysis of {LPG} transport by road and rail. Journal of Loss Prevention in the Process Industries 2000;13:27−31.

[12] Derudi M, Bovolenta D, Busini V, Rota R. Heavy gas dispersion in presence of large obstacles: selection of modeling tools. Industrial and Engineering Chemical Research 2014;53:9303−10.

[13] Busini V, Rota R. Influence of the shape of mitigation barriers on heavy gas dispersion. Journal of Loss Prevention in the Process Industries 2014;29:13−21.

[14] Pontiggia M, Busini V, Derudi M, Alba M, Scaioni M, Rota R, et al. Safety of LPG rail transportation in the perspective of the Viareggio accident. Reliability, Risk, and Safety − Back to the Future 2010: 1872−80.

[15] Courtesy from Ufficio S.I.T − Viareggio Patrimonio srl − Comune di Viareggio, Italy, 2009 [n.d].

[16] Pontiggia M, Derudi M, Alba M, Scaioni M, Rota R. Hazardous gas releases in urban areas: assessment of consequences through CFD modelling. Journal of Hazardous Materials 2010;176:589−96.

[17] DNV. PHAST technical reference manual. v. 6.0. London, UK: DNV; 1999.

[18] Uijt de Haag PAM, Ale BJM. Guidelines for quantitative risk assessment (Purple Book). The Hague (NL): Committee for the Prevention of Disasters; 1999.

CHAPTER 11

Assessing the Severity of Runaway Reactions

V. Casson Moreno[1], E. Salzano[1], F. Khan[2]

[1]University of Bologna, Bologna, Italy; [2]Memorial University of Newfoundland, St John's, NL, Canada

1. INTRODUCTION

Chemical and process industries are historically perceived as unsafe by public opinion. This negative reputation was surely reinforced after the dramatic events that occurred in the past century in Seveso, Italy, and Bhopal, India, two milestones in chemical safety at an industrial scale [1]. These accidents were due to runaway reactions (also called thermal explosions), which are the result of the loss of thermal control in a vessel undergoing a strong exothermic process [2]. Runaway reactions are typically characterized by an exponential increase of the temperature inside the vessel [3] so that the rate of heat generation becomes faster than the rate of heat removal/loss, with a consequent accumulation of heat and acceleration of the reaction rate [4]. This event is more frequent for batch and semibatch chemical processes, in which heat accumulation is more likely to occur [5]. However, thermal runaways might happen any time reactive chemicals are handled, for example, during transportation, storage, or other industrial operations [6,7]. The number of fatalities and injuries related to runaway reactions has surprisingly increased more than three times in the past 30 years, although the total number of events has significantly lowered [8].

For this reason, this chapter addresses fundamental concepts and the typical methodology adopted for the risk analysis related to chemical reaction hazards in the process industry. Furthermore, new methodologies for dynamic risk assessment based on minimum sets of data are introduced.

2. SAFETY MEASURES FOR RUNAWAY REACTIONS

A thermal explosion may lead to catastrophic consequences and dramatically involve the population and the environment. The bow-tie diagram described in Fig. 11.1 depicts the potential scenarios related to runaway reactions. The diagram is generic and only for illustrative purposes. For this reason, bow-tie diagrams reflecting the characteristics of the specific reactions should be defined for each singular analysis to consider the intrinsic hazards of the involved substances (reagents, intermediates, products, and by-products).

Dynamic Risk Analysis in the Chemical and Petroleum Industry
ISBN 978-0-12-803765-2

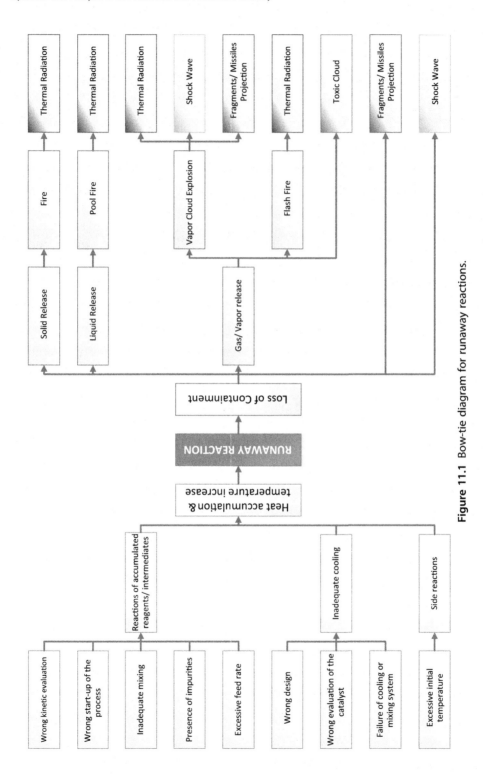

Figure 11.1 Bow-tie diagram for runaway reactions.

In Fig. 11.1, the top event is represented by the runaway reaction event, that is, the moment in which the control of the intrinsic hazard is lost. The left side of the diagram describes the potential causes of a runaway reaction, as follows:

- reactions of accumulated reagents or intermediates due to unexpected reaction kinetics, failures in process start-up, ineffective mixing of reagents inside the reactor, impurities compromising the reaction, or excessive feed rate;
- inadequate cooling, inducing temperature increase in the reactor; can be due to ineffective design or technical failure of the cooling system or unexpected behavior of the catalyst during the reaction;
- side reactions occurring in the reactor as a result of the wrong initial temperature setting.

The right side of Fig. 11.1 outlines the potential consequences of a runaway reaction on humans. A runaway reaction may directly lead to loss of containment (small breach, large breach, or catastrophic rupture). Depending on the type of loss of containment, a shock wave or fragment/missile projection may occur. The type of substances and reactions involved and the dynamics of the loss of containment may lead to different types of release, as follows:

- solid release, which may lead to fire and thermal radiation to humans, if flammable substances are liberated;
- liquid release, which may lead to pool fire and thermal radiation to humans, if flammable substances are liberated;
- gas/vapor release, which may lead to
- vapor cloud explosion (thermal radiation, shock wave, and fragment/missile projection) and flash fire (thermal radiation), if flammable substances are liberated;
- toxic cloud, if toxic substances are liberated.

Two-phase releases are also possible but are not considered in this generic bow-tie diagram.

In this framework, it is clear that specific risk assessment should be carried out for any chemical process if runaway reactions are likely, and appropriate safety measures should be selected, implemented, and maintained. These measures include (1) the reduction of the hazards by inherently safer design, (2) the prevention of the runaway, and (3) the mitigation of the consequences of a runaway [9]. Some of these are completely protective, other are partially preventive [10].

The application of inherent safety principles and inherently safer design concepts has been proven to reduce the risk of accidents for process plants [11]. However, an inherently safer design could possibly alter the hazards in other dimensions [12]. Therefore, this limitation should be overcome before defining an inherently safer design alternative, and the hazards associated with a particular material or process should be thoroughly understood [13].

Active strategies include control systems and safety-instrumented systems. They can either prevent a runaway or minimize the consequences (ie, the early warning detection system coupled with the chemical inhibition method). The prevention of a runaway reaction and its early detection have been widely investigated in the past 20 years. Much literature on thermal explosion theories is available elsewhere [4,14—16].

Reaction inhibition is another active measure involving injection into the reactor of small quantities of reaction inhibitor, decreasing the rate of the runaway at an early stage. Despite the apparent suitability, inhibition systems are rarely used in industrial processes and are not widely studied because of the intrinsic danger associated with the experimental activity [17].

Mitigation of runaway reaction consequences by venting, containment, and venting with containment are passive protective measures aimed at loss prevention. In particular, reactor venting is widely used in industry [18]. The design of protective devices depends on the nature of the system under runaway as well as the reliability of the devices [19]. For instance, there are cases when emergency pressure relief systems might not be economically feasible, such as some batch polymerization reactions [20] generating gas at a very fast rate on decomposition [21], which would need exceptionally large relief valves. For such cases, other strategies, such as prevention, quenching, inhibition, and dumping, are used [14,15,22]. The support given by the layer of protection analysis (and bow-tie analyses inclusive of safety barriers) is also used to determine whether the likelihood of runaway may be reduced by additional safety measures (layers of protection/safety barriers), so that the runaway scenario is no longer is considered possible [23].

3. ANALYSIS OF RISKS RELATED TO RUNAWAY REACTIONS

Most of the safety assessment phases addressing chemical reactions (eg, identification and assessment of related causes and consequences as well as identification and sizing of related preventative and protective measures) are based on experimental activity [1,2,10]. In particular, prediction of the consequences of a runaway reaction, based on temperature and pressure evolution in a reactor, requires the knowledge of the reaction kinetics, thermodynamics, and fluid dynamics inside the vessel. Such phenomena and their interaction are complex and yet need to be fully understood, especially in the case of reactions with pressure generation totally or partially due to production of permanent gases (gassy or hybrid systems) [10,24]. Moreover, these phenomena cannot be easily determined by laboratory-scale experiments [9,21,25]. A vast amount of literature is dedicated to the calorimetric approach for the assessment of thermal hazards [26]. The main features and characteristics of the standard experimental techniques are summarized in Table 11.1. (Related descriptions and definitions are in Tables 11.2—11.4.) On the one hand, the application of a calorimetric approach to risk analysis provides useful data to help predict the behavior of a process: It allows us to judge the suitability of the designed process to

Table 11.1 Comparison of Methods and Techniques

	Thermometry		Calorimetry			
	Differential Thermal Analysis	Nondifferential Thermal Analysis	Differential Scanning Calorimetry	Microcalorimetry	Reaction Calorimetry	Adiabatic Calorimetry
Operating Mode						
Scanning	✓	✓	✓	✓	(✓)	
Isothermal	✓	✓	✓	✓	✓	
Isoperibolic					✓	
Adiabatic					(✓)	✓
Main Measurements						
Temperature	✓	✓				✓
Temperature difference					✓	
Heat released			✓	✓	✓	
Pressure		✓		✓	✓ (✓)	✓
Typical Features						
Sample size (g)	10^{-3}		10^{-3}		10^2–10^3	10
Sensitivity	10^{-6} K	10^{-3} K	10^{-6} W	$<10^{-6}$ W	W	10^{-3} W
Temperature operating range (°C)	−200 to 2000	Ambient to 500	−200 to 600	−20 to 300	−80 to 300	Ambient to 500
Hazard Parameters						
Maximum temperature due to the synthesis reaction			M	M	M	D
Adiabatic temperature increase (ΔT_{ad})			C	C	C	D

Continued

Table 11.1 Comparison of Methods and Techniques—cont'd

	Thermometry			Calorimetry		
	Differential Thermal Analysis	Nondifferential Thermal Analysis	Differential Scanning Calorimetry	Microcalorimetry	Reaction Calorimetry	Adiabatic Calorimetry
Onset temperature (T_{onset})	D	D	D	D	D	D
Maximum self-heating rate						
Time to maximum rate						
Induction period						
Kinetic parameters			M	M	M	M
Self-accelerating decomposition temperature			C	C		C
Heat evolved by the reaction (ΔH_R)			D	D		D

✓, common characteristic; (✓), optional characteristic; C, calculated from measurements; D, direct measurement; M, modeling required.
Adapted from Group HW. HarsBook: a technical guide for the assessment of highly reactive chemical systems. DECHEMA; 2002.

Table 11.2 Description of Experimental Techniques [26]

Differential thermal analysis	Differential thermal analysis records any temperature difference between the sample and a reference while they undergo an identical temperature regimen.
Nondifferential thermal analysis	Nondifferential thermal analysis records the temperature change of the sample subjected to a specific temperature regimen. Usually, pseudoadiabatic instruments are used for screening purposes.
Differential scanning calorimetry	Differential scanning calorimetry records the heat transfer by the sample in comparison with a reference while they undergo an identical temperature regimen.
Microcalorimetry	Microcalorimetry records the heat transfer associated with a chemical reaction, reproducing process conditions (eg, stirring and dosing) in small ampules (eg, 3–20 mL).
Reaction calorimetry	Most common method used to simulate the reaction process under industrial conditions and to record the heat exchanged by the system.
Adiabatic calorimetry	Adiabatic calorimetry records the heat transfer associated with a chemical reaction, minimizing heat transfer between the sample and its surroundings.

avoid unexpected side reactions and decompositions of hazardous chemicals, to correctly size protective measures, and to perform a variety of engineering calculations. On the other hand, experimental activity requires highly specialized and expensive instrument. In addition, collecting and interpreting experimental data may be not only time consuming but also extremely challenging.

In past years, researchers have focused on other types of approaches to assess the risk of thermal explosions aimed at reducing the number of required laboratory-scale experiments and generalizing the results. In particular, the use of models and software for prediction of thermochemical and hazard parameters (ie, whether a reaction is thermodynamically favored, the amount of heat evolved, the adiabatic temperature rise, the stability of chemicals, etc.) has been investigated to support prediction of thermal hazard [27]. However, these types of approaches are not aimed at avoiding experimental activities but at optimizing them. Studies have been devoted to determining the necessary amount of experimental information required for full identification of the complexity of a process [28,29]. Other approaches are focused on risk assessment and are based on semiempirical evaluations, in which a minimum set of experimental data is used to define

Table 11.3 Definitions of Operating Modes [26]

Scanning	A constant heating rate is applied to the sample. This mode is used to obtain reaction kinetics. The results obtained should be validated with isothermal/adiabatic tests.
Isothermal	A constant temperature is applied to the sample. The shape of the experimental curves obtained through this mode gives direct information about the kinetics of the reaction.
Isoperibolic	A constant temperature is applied to the surrounding (reactor) jacket. Its applications are usually limited to simulations of industrial procedures.
(Quasi) adiabatic	Heat transfer between the sample and its surroundings is minimized. This is the most appropriate mode for studying runaway reactions.

Table 11.4 Definitions of Hazard Parameters [26]

Maximum temperature due to the synthesis reaction	Maximum temperature that can be reached owing to the studied reaction under adiabatic conditions. For a batch process, this can be determined by adiabatic calorimetry.
Adiabatic temperature increase (ΔT_{ad})	Adiabatic increase of temperature due to the reactions taking place in the system under analysis.
Onset temperature (T_{onset})	Lowest temperature at which exothermic reaction is observed; it depends on the sensitivity of the instrument used.
Maximum self-heating rate	Maximum rate of increase of temperature under adiabatic conditions.
Time to maximum rate	Time to achieve the maximum self-heating rate under adiabatic conditions (due to side reactions or decomposition). It depends on initial temperature. It should be determined at the maximum temperature due to the synthesis reaction.
Induction period	The time after which exothermic activity will appear under isothermal conditions. The induction time is a function of temperature, concentrations of catalysts/inhibitors, and presence of impurities.
Kinetic parameters	Parameters of reaction kinetics studying the rates of chemical processes.
Self-accelerating decomposition temperature	The temperature of a cooling medium that causes the heat generation rate of the system to be equal to the heat removal rate of the coolant. It depends on the size of container and the thermokinetic parameters of the reaction.
Heat evolved by the reaction (ΔH_R)	Heat produced by an exothermic reaction.

probability and severity of the runaway reaction [30,31]. Minimum data collection leads to decreasing costs of analysis and allows a dynamic approach to risk assessment of chemical reaction hazards. In particular, probability assessment may have the potential to be updated once new sets of data are available [30] or assessed on a real-time basis by means of the current reactor conditions, as described in Chapter 12.

4. DYNAMIC METHOD

The dynamic method proposed is based on a recent study carried out by Wang et al. [31]. This study is aimed at ranking thermal risk on the basis of a kinetic reaction model and a minimum set of parameters such as the onset temperature (T_{onset}), the adiabatic time to maximum rate, and the heat evolved by the reaction (ΔH_R) (see Table 11.4). The method is aimed at measuring online the risk associated with a thermal hazardous reaction by monitoring the temperature inside the reactor. The evaluation is based on a dynamic calculation of the time to maximum rate.

For a decomposition reaction with relatively high activation energy, it has been shown that the adiabatic time to maximum rate can be approximated as [32]

$$\text{TMR}_{\text{ad}} = \frac{c_p \cdot R \cdot T^2}{\dot{q} \cdot E_a} \tag{11.1}$$

where c_p is the specific heat of the reaction mass (J/kg K), R is the gas constant (J/mol K), T is the temperature inside the reactor (K), $\dot{q}$ is the rate of head production of the reaction at any temperature (W/kg), and E_a is the activation energy of the reaction (J/mol). With this approach, (1) the system is assumed to be adiabatic (ie, heat losses are negligible), and (2) the concentration decrease is neglected so that the time to maximum rate calculated is even shorter than the true adiabatic value, which is conservative.

The rate of head production of the reaction $\dot{q}$ is

$$\dot{q} = (-\Delta H_R) \cdot (-r \cdot V) \tag{11.2}$$

where r is the reaction rate (mol/s) and V the reaction volume (m^3). Eq. (11.2) can be estimated from the heat produced by the reaction at the onset temperature ($\dot{q}_0$) as follows:

$$\dot{q} = \dot{q}_0 \cdot \exp\left(\frac{E_a}{R} \cdot \left(\frac{1}{T_0} - \frac{1}{T}\right)\right) \tag{11.3}$$

The expression for energy balance of an adiabatic system is:

$$m \cdot c_p \cdot \frac{dT}{dt} = (-\Delta H_R) \cdot (-r \cdot V) \tag{11.4}$$

Combining Eqs. (11.2)–(11.4), it is possible to relate $\dot{q}_0$ to the temperature increase inside the reaction dT/dt:

$$\dot{q}_0 = \frac{dT}{dt} \cdot c_P \tag{11.5}$$

Lastly, combining Eqs. (11.1), (11.3), and (11.5), the adiabatic time to maximum rate is obtained:

$$\text{TMR}_{ad} = \frac{R \cdot T^2}{E_a \cdot \left(\frac{dT}{dt}\right) \cdot \exp\left(\frac{E_a}{R} \cdot \left(\frac{1}{T_0} - \frac{1}{T}\right)\right)} \tag{11.6}$$

This value can be calculated online if the measurement of temperature for the reacting system is available. The temperature rate can be obtained through the temperature logs. This is a standard parameter monitored for potential thermal runaway [2].

Online evaluation of risk is also possible by means of a dynamic calculation of the thermal risk index (TRI), as stated by Wang et al. [31]. This risk is a combination of severity and probability, and to represent it, severity and probability have to be defined for the reaction under analysis. Severity of a hazardous material is related to the heat that can be released during the associated runaway reaction. Probability is defined on the basis of the runaway time scale and is related to the adiabatic time at maximum rate. To standardize the TRI [31], a reference hazardous material has to be used; its choice needs to be congruent with the type of reaction under analysis (eg, decomposition of organic peroxide or of azo compounds, oxidation of amine compounds, etc.). Finally, TRI is ranked against a risk scale based on the evaluation of the TRI for a pool of hazardous material with respect to a reference.

5. CONCLUSIONS

Risk associated with runaway reactions cannot be disregarded when assessing a chemical plant. Specific safety measures are available and may allow prevention and mitigation of such hazardous events. To improve safety measures, detailed and reliable risk assessment should be carried out. Loss of temperature control is one of the major reasons leading to runaway reaction. To dynamically assess the risk related to a thermal runaway, a method based on direct measures of temperature inside the reactor is proposed.

REFERENCES

[1] Mannan S. Lees' loss prevention in the process industries: hazard identification, assessment and control. 4th ed. Oxford (UK): Elsevier; 2012.
[2] Stoessel F. Thermal safety of chemical processes. Weinheim; 2008.
[3] Crowl DA, Louvar JF, Daniel A, Crowl JFL. Chemical process safety. 2011.

[4]   Varma A, Morbidelli M, Wu H. Parametric sensitivity in chemical systems. Cambridge (UK): Cambridge University Press; 1999.

[5]   Copelli S, Torretta V, Pasturenzi C, Derudi M, Cattaneo CS, Rota R. On the divergence criterion for runaway detection: application to complex controlled systems. Journal of Loss Prevention in the Process Industries 2014;28:92−100. http://dx.doi.org/10.1016/j.jlp.2013.05.004.

[6]   Vilchez J, Sevilla S, Montiel H, Casal J. Historical analysis of accidents in chemical plants and in the transportation of hazardous materials. Journal of Loss Prevention in the Process Industries 1995;8: 87−96.

[7]   Casson V, Maschio G. Risk analysis in transport and storage of monomers : an accident investigation. Macromolecular Symposia 2011;302:273−9. http://dx.doi.org/10.1002/masy.201000068.

[8]   Saada R, Patel D, Saha B. Causes and consequences of thermal runaway incidents—will they ever be avoided? Process Safety and Environmental Protection 2015. http://dx.doi.org/10.1016/j.psep.2015.02.005.

[9]   Casson Moreno V, Kanes R, Wilday J, Véchot L. Modeling of the venting of an untempered system under runaway conditions. Journal of Loss Prevention in the Process Industries 2015;36:171−82. http://dx.doi.org/10.1016/j.jlp.2015.04.016.

[10] Center for Chemical Process Safety (CCPS). Guidelines for pressure relief and effluent handling systems. 1998.

[11] Rusli R, Shariff AM, Khan FI. Evaluating hazard conflicts using inherently safer design concept. Safety Science 2013;53:61−72. http://dx.doi.org/10.1016/j.ssci.2012.09.002.

[12] Center for Chemical Process Safety (CCPS). Guidelines for risk based process safety. AIChe; 2007.

[13] Center for Chemical Process Safety (CCPS). Inherently safer chemical processes: a life cycle approach. 2nd ed. Wiley; 2008.

[14] Morbidelli M, Varma A. A generalized criterion for parametric sensitivity: application to thermal explosion theory. Chemical Engineering Science 1988;43:91−102.

[15] Casson V, Lister DG, Milazzo MF, Maschio G. Comparison of criteria for prediction of runaway reactions in the sulphuric acid catalyzed esterification of acetic anhydride and methanol. Journal of Loss Prevention in the Process Industries 2012;25:209−17.

[16] Maestri F, Copelli S, Rota R, Gigante L, Lunghi A, Cardillo P. Simple procedure for optimal scale-up of fine chemical processes. II. Nitration of 4-chlorobenzotrifluoride. Industrial & Engineering Chemistry Research 2009;48:1316−24. http://dx.doi.org/10.1021/ie800466n.

[17] Dakshinamoorthy D, Louvar JF. Shortstopping and jet mixers in preventing runaway reactions. Chemical Engineering Science 2008;63:2283−93. http://dx.doi.org/10.1016/j.ces.2007.05.008.

[18] Rowe SM, Nolan PF, Starkie AJ. Control of runaway polymerization reactions by inhibition techniques. Inst chem eng symp ser, vol. 134. Inst Chem Eng; 1994. p. 575−87.

[19] Stoessel F. Planning protection measures against runaway reactions using criticality classes. Process Safety and Environmental Protection 2009;87:105−12. http://dx.doi.org/10.1016/j.psep.2008.08.003.

[20] Casson V, Snee T, Maschio G. Investigation of an accident in a resins manufacturing site: the role of accelerator on polymerisation of methyl methacrylate. Journal of Hazardous Materials 2014;270: 45−52. http://dx.doi.org/10.1016/j.jhazmat.2014.01.038.

[21] Reyes Valdes OJ, Casson Moreno V, Waldram SP, Véchot LN, Mannan MS. Experimental sensitivity analysis of the runaway severity of Dicumyl peroxide decomposition using adiabatic calorimetry. Thermochimica Acta 2015;617:28−37. http://dx.doi.org/10.1016/j.tca.2015.07.016.

[22] Copelli S, Derudi M, Rota R. Topological criterion to safely optimize hazardous chemical processes involving arbitrary kinetic schemes. Industrial & Engineering Chemistry Research 2011;50:1588−98. http://dx.doi.org/10.1021/ie102014w.

[23] Center for Chemical Process Safety (CCPS). Layer of protection analysis: simplified process risk assessment. John Wiley & Sons; 2011.

[24] Etchells J, Wilday J. Workbook for chemical reactor relief system sizing. Crown; 1998.

[25] Snee TJ, Cusco L. Pilot-scale evaluation of the inhibition of exothermic runaway. Process Safety and Environmental Protection 2005;83:135−44. http://dx.doi.org/10.1205/psep.04240.

[26] Group HW. HarsBook: a technical guide for the assessment of highly reactive chemical systems. DECHEMA; 2002.

[27] Pasturenzi C, Dellavedova M, Gigante L, Lunghi A, Canavese M, Cattaneo CS, et al. Thermochemical stability: a comparison between experimental and predicted data. Journal of Loss Prevention in the Process Industries 2014;28:79–91. http://dx.doi.org/10.1016/j.jlp.2013.03.011.

[28] Roduit B, Borgeat C, Berger B, Folly P, Alonso B, Aebischer JN. The prediction of thermal stability of self-reactive chemicals. Journal of Thermal Analysis and Calorimetry 2005;80:91–102. http://dx.doi.org/10.1007/s10973-005-0619-4.

[29] Sanchirico R, Tecchio PV. Model selection and parameters estimation in kinetic thermal evaluations using semiempirical models. AIChE Journal 2012;58:1869–79. http://dx.doi.org/10.1002/aic.

[30] Busura S, Khan F, Hawboldt K, Iliyas A. Quantitative risk-based ranking of chemicals considering hazardous thermal reactions. Journal of Chemical Health and Safety 2014;21:27–38. http://dx.doi.org/10.1016/j.jchas.2014.03.001.

[31] Wang Q, Rogers WJ, Mannan MS. Thermal risk assessment and rankings for reaction hazards in process safety. Journal of Thermal Analysis and Calorimetry 2009;98:225–33. http://dx.doi.org/10.1007/s10973-009-0135-z.

[32] Frank-Kamenetskii DA. Diffusion and heat transfer in chemical kinetics. 2nd ed. New York, NY: Plenum Press; 1969.

CHAPTER 12

Dynamic Assessment of Runaway Reaction Risk: Tutorial and Examples

V. Casson Moreno[1], E. Salzano[1], N. Paltrinieri[2,3]

[1]University of Bologna, Bologna, Italy; [2]Norwegian University of Science and Technology (NTNU), Trondheim, Norway; [3]SINTEF Technology and Society, Trondheim, Norway

1. INTRODUCTION

A tutorial for the methodology introduced in the previous chapter is presented. This technique is also applied to the assessment of thermal hazards associated with the decomposition reaction of dicumyl peroxide (DCP) under runaway conditions [1].

This organic peroxide is commonly used in the polymer industry as a curing agent for unsaturated polystyrene and as a crosslinking agent for polyethylene, ethylene vinyl acetate copolymers, and others. It is also added to different resins to improve the physical properties of electronic components, footwear, electrical insulators, surface coatings, and architectural materials. However, because of the instability of the oxygen–oxygen bond, DCP can exothermically self-decompose, leading to a runaway reaction accompanied by a rapid pressure increase; several recent accidents testify to this, for example, Taiwan (1988, 2003, and 2008) and Japan (1999) [1]. Because of this, DCP was chosen for the application of the methodology for dynamic risk assessment of thermal hazards introduced in the previous chapter.

From a theoretical point of view, a large-scale vessel under runaway condition behaves like an adiabatic system [2–4] because the heat produced by the reaction is increasingly accumulated in the system. For this reason, an adiabatic calorimetry experiment is the best experimental option [2], even if it is the most expensive, time consuming, and difficult-to-handle technique. The case study described in this chapter is based on adiabatic calorimetry experiments that allowed safe online monitoring of the temperature profile inside the vessel during the runaway decomposition at a laboratory scale. A schematic of the configuration used to apply the methodology for the dynamic risk assessment of thermal hazards is depicted in Fig. 12.1.

2. METHODOLOGY TUTORIAL

Let us consider a reactor carrying on a reaction/process that might undergo a runaway path. Six main steps preceded by a set of prerequirements (Fig. 12.2) should be performed to apply the proposed methodology for dynamic assessment of runaway reaction risk.

Dynamic Risk Analysis in the Chemical and Petroleum Industry
ISBN 978-0-12-803765-2
139

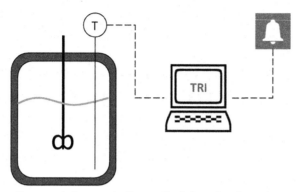

Figure 12.1 Configuration needed to apply the methodology for dynamic assessment of runaway reaction risk.

2.1 Prerequirements

The following are the necessary prerequirements:

- Monitoring the temperature of the reactor content during the reaction. A data acquisition system needs to be in place and the acquired data have to be logged into a calculator to perform thermal risk index (TRI) calculations.
- The activation energy of the reaction (E_a) has to be defined. It can be previously determined using any of the techniques mentioned in the previous chapter.
- Reference hazardous material and risk scale need to be set. The former is used to calculate the TRI, the latter to compare its value and rank the risk associated with the reaction under analysis.

The choice of the reference hazardous material is related to the reaction under study. In regard to components of the risk scale, Wang et al. [5] determined one based on the evaluation of the TRI of a pool of hazardous materials for which data were available in the literature. In this chapter, the same scale was adopted (Table 12.1).

2.2 Steps 1 and 2: Normal Operating Conditions

During the course of a reaction, the temperature of the reacting mass (T, step 1) is monitored and logged into a computer that performs the calculation of the temperature rate (dT/dt, step 2). Details on how to perform good data acquisition and an efficient and accurate evaluation of temperature rate can be found elsewhere [6–10].

If the rate of temperature becomes higher than a threshold value, the reaction moves from normal operating conditions to self-heating mode (eg, a runaway reaction is occurring). The choice of the threshold value depends on the sensitivity of the equipment [2]: The lower the sensitivity, the lower the detected onset temperature for the runaway reaction. Once the threshold is passed, the data acquisition system has to log the onset temperature of the reaction (T_0).

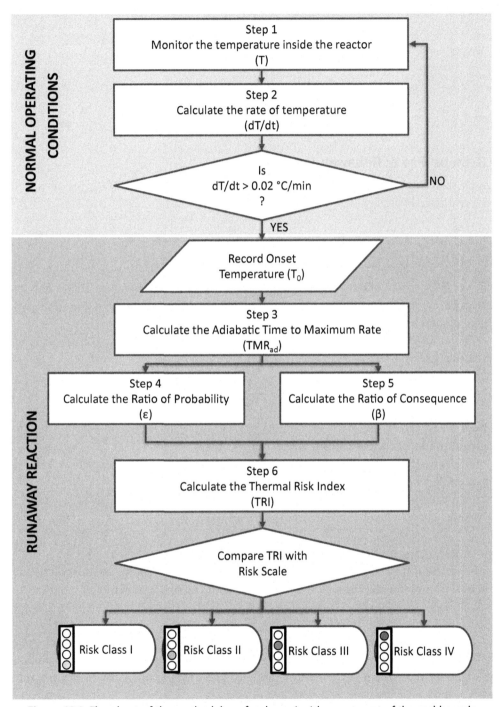

Figure 12.2 Flowsheet of the methodology for dynamic risk assessment of thermal hazards.

Table 12.1 Risk scale adopted in this chapter [5]

Risk class	Thermal risk index (TRI) value
I	$TRI < 1$
II	$1 < TRI < 2$
III	$2 \leq TRI < 3$
IV	$TRI \geq 3$

2.3 Steps 3 to 6: Runaway Reaction

Under runaway conditions, the adiabatic time to maximum rate (TMR$_{ad}$) is calculated on a real-time basis (step 3)

$$TMR_{ad} = \frac{R \cdot T^2}{E_a \cdot \left(\frac{dT}{dt}\right) \cdot \exp\left(\frac{E_a}{R} \cdot \left(\frac{1}{T_0} - \frac{1}{T}\right)\right)} \tag{12.1}$$

where R = gas constant (J/mol K); T = absolute temperature (K); E_a = activation energy (J/mol); t = time (min); T = absolute temperature (K); and T_0 = onset temperature of reaction (K).

The ratio of probability (ε) is obtained from the real-time TMR$_{ad}$ of the reaction under analysis (step 4)

$$\varepsilon = \frac{TMR_{ad,ref}}{TMR_{ad}} \tag{Eq. 12.2}$$

where TMR$_{ad,ref}$ = time to maximum rate of the reference compound used to standardize the risk assessment method.

The ratio of severity (β) is obtained from the characteristic heat of reaction ΔH_R of the reaction under analysis (step 5)

$$\beta = \frac{\Delta H_R}{\Delta H_{R,DTB}} \tag{Eq. 12.3}$$

where $\Delta H_{R,ref}$ = heat of reaction of the reference compound.

In step 6, TRI is calculated on a real-time basis from the ratios of probability and severity, to monitor the thermal risk associated with the ongoing runaway reaction.

$$TRI = \varepsilon \times \beta \tag{Eq. 12.4}$$

TRI is also compared with the predefined risk classes to define different preventative and protective measures to implement, as described in the previous chapter (eg, addition of inhibitors, blowdown of the reactor contents, and opening of an emergency pressure relief system).

3. RESULTS

The application of the methodology is shown for the analysis of thermal decomposition of DCP 30% by weight in cumene. The runaway experiments were performed using a Phi-TEC II (HEL, Borehamwood, UK) adiabatic calorimeter [1], which allows one to monitor online the temperature of the reaction and calculate the related rate of temperature. In this experimental apparatus, the hazardous material is driven to runaway in a 110-mL thin-walled stirred test cell. Phi-TEC II has excellent adiabatic performances owing to the relatively low thermal mass of the sample holder compared with the thermal mass of the substance under reaction. This ensures that experimental data can closely predict large-scale behavior [2]. The equipment can work up to 500°C and 120 bars. The lowest possible exotherm detection threshold is 0.02°C/min. For this reason, this value is the set threshold value used to detect T_0.

In regard to the choice of the reference compound, di-tert-butyl peroxide is chosen because it has been extensively studied, and its runaway behavior is well known [5,11]. This choice is corroborated by the fact that DCP is an organic peroxide as well undergoing a decomposition. This defines the values of $\Delta H_{R,ref}$ and $TMR_{ad,ref}$ as reported in Table 12.2.

The activation energy of the solution tested is retrieved from the literature [1] ($E_a = 154$ kJ/mol).

Fig. 12.3 shows the temperature and rate of temperature versus time monitored during the runaway decomposition of DCP. It is possible to appreciate that the onset temperature of the reaction was detected at 115°C.

A specific focus on the profiles during the runaway reaction (ie, from onset to maximum temperature) is shown in Fig. 12.4.

From the onset temperature onward, the adiabatic time to maximum rate (TMR_{ad}) and TRI are calculated. The TMR_{ad} profile versus time is displayed in Fig. 12.5 in comparison with the temperature profile of the sample. It is possible to notice its rapid decreasing trend linked to an increasing self-heating of the system. Being calculated on the basis of the rate of temperature, the profile of TMR_{ad} is affected by noise in the temperature signal, as is dT/dt (see Fig. 12.4). Specific methods for noise reduction are reported elsewhere [6–10].

TRI is calculated at every temperature data log, obtaining the profile versus time shown in comparison with the temperature profile in Fig. 12.6.

Table 12.2 Reference compound properties [5]
Di-tert-butyl peroxide

$\Delta H_{R,ref}$	556 [J/g]
$TMR_{ad,ref}$	98 [min]

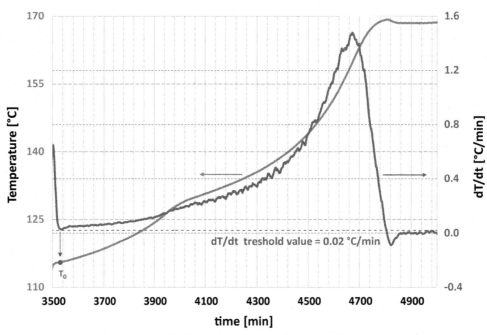

Figure 12.3 Temperature (*left axis*) and rate of temperature (*right axis*) versus time.

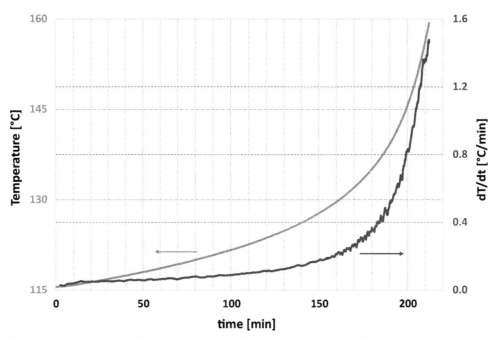

Figure 12.4 Temperature (*left axis*) and rate of temperature (*right axis*) versus time during the runaway decomposition.

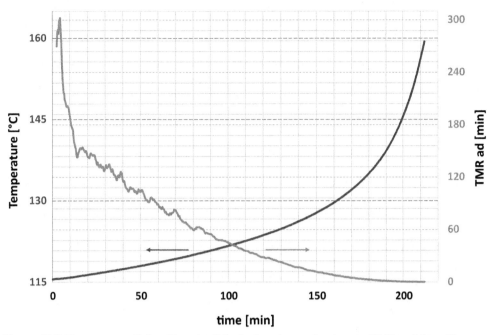

Figure 12.5 Temperature (*left axis*) and adiabatic time to maximum rate (TMR$_{ad}$, *right axis*) versus time.

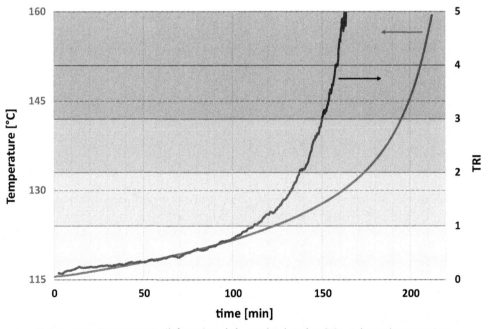

Figure 12.6 Temperature (*left axis*) and thermal risk index (TRI, *right axis*) versus time.

As per the previous discussion, TRI calculation is affected by the noise of the temperature signal, but it is not relevant for our analysis, which is focused on a ranking with respect to a predefined risk scale. In fact, Fig. 12.6 only reports TRI values that are comparable to the predefined risk scale (see Table 12.1). A choice of colors for the scale (assigned to increasing TRI values) helps in the visualization of the ranking, but this is at the analyst's discretion.

4. DISCUSSION

Fig. 12.7 reports TMR_{ad} (*left axis*) and TRI (*right axis*) profiles versus time. The respective values are shown in the predefined risk scale range. The colors assigned to the increasing risk profile may support the visualization of TRI, but they are at the analyst's discretion.

The value of TMR_{ad} at the beginning of the runaway reaction (ie, at the onset temperature) was nearly 5 h. This timespan allows the implementation of corrective measures such as the injection of inhibitors or of a cooling agent. Accordingly, the reaction is classified in risk class I.

In less than 20 min, TMR_{ad} decreases by half because of the strong self-heating behavior of DCP. The corresponding value of TRI ranks the process in risk class II. In this case, it might be possible to take actions such as discharging the reactor content (if possible and allowed by company procedures).

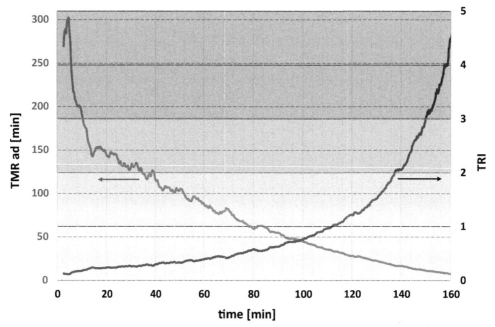

Figure 12.7 Adiabatic time to maximum rate (TMR$_{ad}$, *left axis*) and thermal risk index (TRI, *right axis*) versus time.

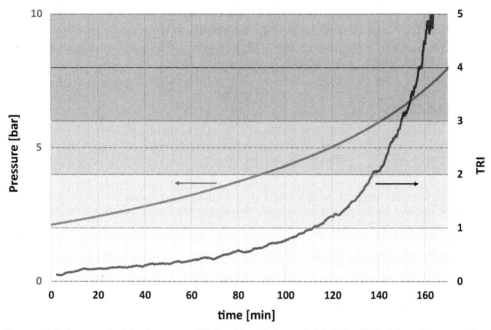

Figure 12.8 Pressure inside the test cell (*left axis*) and thermal risk index (TRI, *right axis*) versus time.

The process enters risk class III when there is less than 30 min to the maximum rate of temperature. This timespan starts to be very short to implement sophisticated measures, such as those mentioned previously.

When risk class IV is reached, only 10 min are left to reach the maximum rate of temperature inside the reactor. This TMR_{ad} allows only drastic emergency measures, such as the evacuation of the site. In the specific case of the thermal decomposition of DCP, when the reaction is in risk class IV, the pressure inside the test cell reached nearly 6 bars (Fig. 12.8). Unless adequate protective measures (emergency pressure relief system [3,4,12]) are in place, this pressure could be sufficient to burst the reactor.

The use of a TRI and related risk classes conveys additional information on potential runaway consequences. In fact, TRI includes the ratio of severity β and allows prioritizing high-impact reactions while planning for risk prevention/mitigation.

5. CONCLUSIONS

The proposed method allows the drawing of useful recommendations regarding safe operations related to hazardous materials and chemical processes. Thermal risk associated with a hazardous material undergoing a runaway reaction can be assessed online while the reaction is taking place. Risk assessment and ranking are possible in real time, allowing the proposition of adequate safety and emergency measures. A minimum set of prerequirements is needed for the application of this methodology. In this way, the methodology may be implemented at an industrial scale with the aim of improvement of process safety, loss prevention, and emergency response for aspects related to chemical reaction hazards and runaway reactions.

REFERENCES

[1] Reyes Valdes OJ, Casson Moreno V, Waldram SP, Véchot LN, Mannan MS. Experimental sensitivity analysis of the runaway severity of dicumyl peroxide decomposition using adiabatic calorimetry. Thermochimica Acta 2015;617:28—37. http://dx.doi.org/10.1016/j.tca.2015.07.016.

[2] HarsNet. HarsBook. Frankfurt: Dechema; 2002.

[3] Etchells J, Wilday J. Workbook for chemical reactor relief system sizing. Crown; 1998.

[4] Casson Moreno V, Kanes R, Wilday J, Véchot L. Modeling of the venting of an untempered system under runaway conditions. Journal of Loss Prevention in the Process Industries 2015;36:171—82. http://dx.doi.org/10.1016/j.jlp.2015.04.016.

[5] Wang Q, Rogers WJ, Mannan MS. Thermal risk assessment and rankings for reaction hazards in process safety. Journal of Thermal Analysis and Calorimetry 2009;98:225—33. http://dx.doi.org/10.1007/s10973-009-0135-z.

[6] Kostelich EJ, Schreiber T. Noise reduction in chaotic time-series data: a survey of common methods. Physical Review E 1993;48:1752—63.

[7] Madden HH. Comments on the Savitzky-Golay convolution method for least-squares fit smoothing and differentiation of digital data. Analytical Chemistry 1978;50:1383—6. http://dx.doi.org/10.1021/ac50031a048.

[8] Press WH, Teukolsky SA, Vetterling WT, Flannery BP. Numerical recipes in C: the art of scientific computing. 2nd ed. Cambridge, New York: Cambridge University Press; 1992.

[9] Savitzky K, Golay MJE. Smoothing and differentiation of data by simplified least squares procedures. Analytical Chemistry 1964;36:1627—39.

[10] Casson V, Salzano E, Maschio G. Sensitivity analysis for the thermal stability criteria of hydrogen peroxide. Chemical Engineering Transactions 2013;32:541—6. http://dx.doi.org/10.3303/CET1332091.

[11] Casson V, Maschio G. Screening analysis for hazard assessment of peroxides decomposition. Industrial and Engineering Chemistry Research 2011:7526—35. http://dx.doi.org/10.1021/ie201690n.

[12] Center for Chemical Process Safety (CCPS). Guidelines for pressure relief and effluent handling systems. 1998.

Establishing the Risk Picture

CHAPTER 13

Risk Metrics and Dynamic Risk Visualization

N.J. Edwin[1,2], N. Paltrinieri[2,3], T. Østerlie[3]
[1]Safetec, Trondheim, Norway; [2]SINTEF Technology and Society, Trondheim, Norway; [3]Norwegian University of Science and Technology (NTNU), Trondheim, Norway

1. INTRODUCTION

Traditional risk metrics may suffer from a series of limitations [1], but one of their main shortcomings is due to their evaluation process, which is often unable to adapt and approximate to ever-changing real-world systems. For this reason, several efforts have been made since 2000 to develop quantitative methods assessing the risk picture variations for chemical and process facilities. One such development has been toward the definition of broad indicator sets for major accident risk in the wake of the Texas City, US, disaster in 2005 [2]. Most of these sets have been initially considered impractical to monitor owing to the high number of indicators [3]. However, the advent of new information and communication technologies has substantially improved the capabilities of online collection and elaboration of such data. For this reason, dynamic visualization of risk analysis results may represent a viable support for safety-critical decision-making and priority definition.

This chapter first addresses the state of the art of traditional risk metrics and the related limitations. Second, a summarized review of dynamic risk visualization solutions is presented to suggest a potential response to such limitations.

2. TRADITIONAL RISK METRICS

Risk metrics represent the main overall output of risk analysis. Johansen and Rausand [1] define a risk metric as a quantitative expression that can be used to answer an aspect of the three questions related to risk:

1. What can happen?
2. How likely is it?
3. What are the consequences?

Risk metrics allow for communication of quantitative risk evaluation and have the potential for effectively supporting safety-critical decision-making. They are employed in different phases of a plant's life cycle for a series of purposes. For instance, risk metric values obtained from risk analysis are compared against criteria from regulations, standards, and best practices (eg, the UK's Health and Safety Executive and the "Purple Book" of The Netherlands [4,5]) to assess whether the system's risk level is under the acceptability/tolerability threshold or whether further risk mitigation actions should be taken. Evaluation of risk metrics may support decisions on design, operational planning, modifications, and specific maintenance strategies.

For this reason, the choice of risk metrics is critical because it directs the kind of information to obtain from risk analysis and whether the results are considered legitimate and informative by decision-makers and stakeholders [6].

Table 13.1 lists a series of representative risk metrics used in the chemical and petroleum industry with their respective formulas and definitions.

Some of the risk metrics listed in Table 13.1 can be visualized in diagrams, such as bar charts, to better communicate their value and compare with potential acceptability thresholds (eg, potential life loss, fatal accident rate). The risk metric "localized individual risk" may be plotted on the site map to outline location-specific risk contours and support land-use planning. Fig. 13.1 shows an example of such plotting obtained for the Buncefield, UK, oil depot (adapted from Ref. [14]).

To highlight the potential for extreme events, societal risk can be plotted on an FN diagram [Frequency (F) of accidents with a number (N) of fatalities], as in Fig. 13.2. Such a diagram represents the relationship between frequency and the number of fatalities in a given population from the realization of specific hazardous events [15].

2.1 Limitations of Traditional Risk Metrics

The highest risk reported in Fig. 13.1 for the Buncefield oil depot is located in the loading gantry area [localized individual risk per annum (LIRA) equal to 1×10^{-4}], whereas the hydrocarbon fuel tanks have the effect of broadening the risk contour associated with LIRA equal to 1×10^{-7}/year. Such an evaluation would presumably lead to prioritization of improvement and maintenance of safety measures to prevent leakage during road tanker loading. However, recent history has shown the criticality of areas for hydrocarbon fueling of tanks, from where catastrophic scenarios may unexpectedly arise: In 2005, the overfilling of a fuel tank at the Buncefield oil depot led to the formation of a flammable cloud, ignition of which caused a series of explosions that destroyed most of the site [14]. Similar accident scenarios have occurred before and after Buncefield in several other oil depots around the world, as reported by Paltrinieri et al. [16].

Fig. 13.1 may have represented misleading support to Buncefield risk management (especially in the days before the accident), and this reflects the main limitation of risk metrics in the evaluation process. In fact, such an evaluation, once it is performed the first time, is hardly ever updated with new evidence/risk-related information, as mentioned in Chapter 1. An industrial system is subject to ever-changing factors (ie, degradation of technical safety equipment and worsening of organizational conditions)

Table 13.1 Representative Risk Metrics

Name	Formula	Definition
Individual risk per annum (IRPA)	$IRPA = \sum_{i=1}^{n} \lambda_i \cdot Pr(E_i) \cdot Pr(D_i|E_i)$	Expected number of individuals killed owing to exposure to hazardous events per year [8]
Localized Individual risk per annum (LIRA)	$LIRA(x,y) = \sum_{i=1}^{n} \lambda_i \cdot Pr(D_i(x,y))$	Expected number of unprotected individuals permanently present at a specified location who were killed owing to exposure to hazardous events per year [8]
Individual risk of dangerous dose (IR_{HSE})	$IR_{HSE,i} = \lambda_{i,10^6} \cdot Pr(D_t \geq DD_t)$	Expected number of cases of individuals receiving a dangerous dose of toxic chemical, which leads to severe distress, injury, or fatality, per 10^6 years [9]
Potential life loss (PLL)	$PLL = np \cdot IRPA$	Expected number of fatalities within a specific population per year [10]
Fatal accident rate (FAR)	$FAR = \frac{PLL \cdot 10^8}{AE}$	Expected number of fatalities within a specific population per 100 million hours of exposure [8]
Weighted risk integral (RI_{COMAH})	$RI_{COMAH} = \sum_{N=1}^{N_{max}} F(N) \cdot N^k$	Expected number of fatalities per year corrected for risk aversion [11]
Total risk (TR)	$TR = PLL + k \cdot \sigma$	Expected number of fatalities within a specific population per year corrected for risk aversion [10]
Potential equivalent fatality (PEF)	$PEF = PLL + 0.1Mj + 0.01Mn$	Expected harm from fatalities, major injuries, and minor injuries per year [7]
Potential environmental risk (PER)	$PER_{EDi} = \lambda_i \cdot Pr(E_i) \cdot Pr(ED_i|E_i)$	Expected number of cases of defined environmental damage for a certain organism, population, habitat, or ecosystem within an area per year [7]
Exceeded recovery time (ERT)	$ERT_{EDi} = \lambda_i \cdot Pr(T_{EDi} > RT_{EDi})$	Expected number of cases of defined environmental damage exceeding the time needed by the ecosystem to recover from damage per year [7]
Economic loss (EL)	$EL = \sum_{i=1}^{n} \lambda_i \cdot C_i$	Expected value of economic loss due to the cost of the number of hazardous events per year [12]
Monetary collective risk (MCR)	$MCR = \sum_{i=1}^{n} \lambda_i \cdot \omega_i \cdot k_i$	Expected value of economic loss calculated by means of willingness to pay for avoiding hazardous events per year and corrected for risk aversion [13]

AE, Number of hours a specific population is exposed to risk in one year; C_i, cost of hazardous event *i*; DD_t, dangerous dose of the toxic chemical *t*; D_t, $D_i(x,y)$, death due to the hazardous event *i*, death due to the hazardous event *i* at the location (x,y); D_t, dose of the toxic chemical *t*; ED_i, environmental damage due to the hazardous event *i*; E_i, exposure to the hazardous event *i*; $F(N)$, yearly frequency of accidents with *N* fatalities; *i*, hazardous event; *k*, k_i, risk aversion, risk aversion to the hazardous event *i*; *Mj*, yearly frequency of major injuries; *Mn*, yearly frequency of minor injuries; *N*, number of fatalities; *n*, number of identified hazardous events; *np*, number of people in the population; Pr, probability; RT_{EDi}, ecosystem recovery time from the environmental damage ED due to the hazardous event *i*; T_{EDi}, duration of the environmental damage ED due to the hazardous event *i*; (x,y), location coordinates; λ_i, yearly frequency of the hazardous event *i*; $\lambda_{i,10^6}$, frequency of the hazardous event *i* per 10^6 years; σ, standard deviation; ω_i, willingness to pay for avoiding hazardous event *i*.
Adapted from Johansen IL, Rausand M. Risk metrics: interpretation and choice. In: Industrial Engineering and Engineering Management (IEEM), IEEE international conference on 2012; 2012. p. 1914–18.

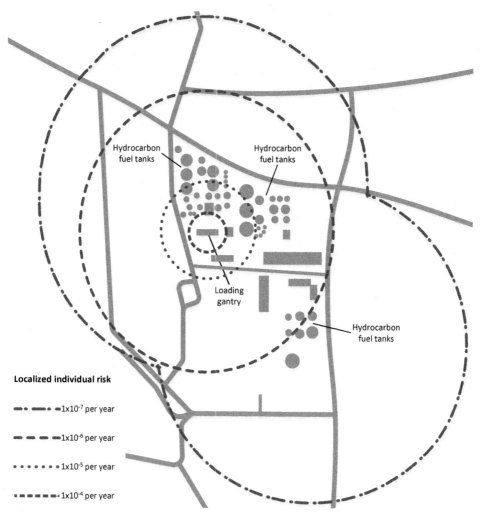

Figure 13.1 Localized individual risk plot of the Buncefield oil depot. *(Adapted from Buncefield Major Incident Investigation Board. The Buncefield incident 11 December 2005: the final report of the major incident investigation board. Bootle (UK): HSE Books; 2008.)*

that may ultimately affect the overall risk. Moreover, risk metrics used for the feasibility/design life cycle phase may prove to be inappropriate for other phases, such as the operational phase [17].

Regarding the Buncefield accident, appropriate technical and organizational indicators might have potentially detected an increasing risk in the system (and updated the risk picture defined by Fig. 13.1). For instance, the automatic gauging system of the overfilled tank had stuck 14 times in the 4 months preceding the accident when the operators were subject to exceptional workload shortly before the accident [18].

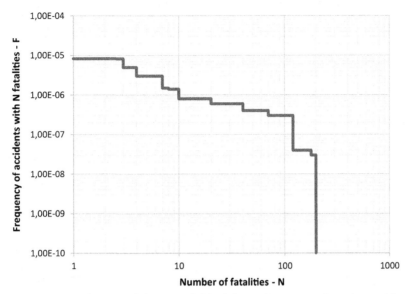

Figure 13.2 Societal risk diagram of the Buncefield oil depot. *(Frequency of accidents with number of fatalities diagram adapted from Buncefield Major Incident Investigation Board. The Buncefield incident 11 December 2005: the final report of the major incident investigation board. Bootle (UK): HSE Books; 2008.)*

As Johansen and Rausand [1] affirm, a challenging topic is the interpretation of uncertainty in relation to risk metrics. Dynamic evaluation consciously addresses such uncertainty by continuously updating and refining risk metrics on the basis of new collected notions.

Another important limitation concerns the aggregation of information to evaluate risk metrics and the inability to distinguish it. No explicit reference to the different affected subsystems (eg, safety barrier systems) is conveyed by the metrics, except for LIRA, which refers to a specific location and may allow for approximate distinction. Subsystems relate to the loading gantry are more critical than subsystems related to the southeast hydrocarbon fuel tanks in Fig. 13.1. However, it must be mentioned that some of the risk metrics listed in Table 13.1 are often calculated for single areas of the system (eg, potential life loss and fatal accident rate) and may allow for the same considerations as LIRA.

Moreover, most of the risk metrics listed in Table 13.1 (except for individual risk of dangerous dose, potential environmental risk, and exceeded recovery time) combine several hazardous events into a single value and may hide differences and relationships of the consequence dimensions considered. The FN diagram in Fig. 13.2 may represent a solution to distinguish between the potential for rare major accidents versus frequent minor incidents. However, as demonstrated by the experience related to the Buncefield accident [14], the concept of societal risk conveyed by the FN diagram may be challenging to understand and implement by potential recipients and authorities. This may be because societal risk is not a single figure that can be displayed as a contour on a map, such as LIRA.

3. DYNAMIC RISK VISUALIZATION

Visualization of risk assessed on a dynamic basis may represent an important support to decision-making and allow one to overcome some of risk metric's limitations. For this reason, in the recent past, operators within the offshore oil and gas sector have developed their own proprietary tools to promote better situational awareness to improve their decision-making [19,20]. These are typically software solutions that support integrated management of safety-critical information through visualization and integrated data management. They provide a coarse qualitative overview of various conditions across the installation.

A step forward from the evolution of these tools is represented by the approach suggested by Hauge et al. [21]: the risk barometer. In this technique, risk-based aggregation rules (calibrated through expert judgment and prior knowledge of system/component importance) are coupled with consolidated risk models (eg, fault tree, event tree, and bow-tie analyses) and used to aggregate the status of indicators for high-level, real-time risk visualization. Examples of risk visualization using the risk barometer approach are discussed in the following section.

3.1 Risk Visualization Elements

This section presents a set of graphical elements used by the risk barometer to visualize different aspects of the risk picture.

3.1.1 Risk Barometer

The risk barometer is used to visualize the real-time risk of a system. It translates risk expressed by means of the metrics described in Section 2 into a relative percentage value indicated by a needle (Fig. 13.3).

Defining and expressing risk solely with a relative measure might be challenging to interpret [22]. However, its use along with a visualization medium (such as the barometer) can mitigate the challenges of misinterpretation of a relative measurement. Use of the barometer for risk visualization involves the definition of a fixed percentage scale with levels of risk tolerability/acceptability expressed through color or section divisions on the barometer. The definition and associated calibration are not fixed but may be based on a number of issues, as follows:

- purpose for which the risk barometer is used (eg, support to long-term versus short-term maintenance planning)
- needs of the user and individual company policies (eg, some companies avoid the association of red with risk)

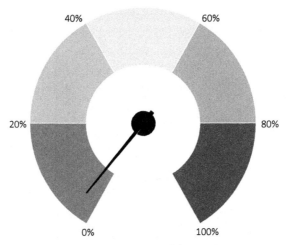

Figure 13.3 Generic risk barometer.

- risk acceptance criteria from regulations, standards, and best practice (eg, as low as reasonably practicable) risk values used in the United Kingdom and the Netherlands [4,5])
- the risk model on which the risk barometer is based, which allows for the assessment of the overall risk (eg, through the use of semiqualitative aggregation rules or quantitative risk assessment)

Definition and choice of the related colors are at the analyst's discretion and should be addressed after an accurate assessment of the aforementioned and other related issues.

In particular, the risk model used for the assessment of the overall risk may reveal the complexity of the system under study and challenge the definition of the levels of risk tolerability/acceptability. For instance, Fig. 13.4 shows an example based on the study carried out by Paltrinieri et al. [23] on the process area of an oil platform. In this case, the overall system risk (represented by the potential life loss) does not vary linearly with the health of the single safety barriers but instead increases exponentially. For this reason, if the barometer is set with risk tolerability/acceptability levels focusing on the health of single safety barriers, the result would correspond to the risk barometer shown in Fig. 13.4. Addressing such issues mitigates the drawbacks of using a relative risk measure because other aspects such as the absolute risk value associated with the relative measurement are accounted for through barometer calibration.

3.1.2 Risk Trend

In addition to the barometer's displaying real-time risk, the risk trend over time can be visualized to evaluate positive or negative trends and compare the current risk with past

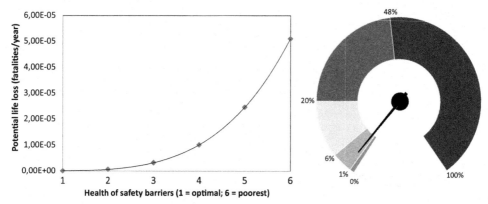

Figure 13.4 Plot of the overall system risk (expressed as potential life loss) in the function of the health of system safety barriers and risk barometer with risk tolerability/acceptability levels defined according to the health of safety barriers. In the *dark green* (dark gray in print versions) level all the barriers have health between 1 and 2; in the *light green* (light gray in print versions) level health is between 2 and 3; in the *yellow* (very light gray in print versions) level health is between 3 and 4; in the *orange* (gray in print versions) level health is between 4 and 5; and in the *red* (very dark gray in print versions) level health is between 5 and 6. This example is based on the study carried out by Paltrinieri et al. [23].

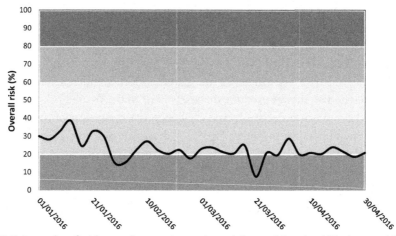

Figure 13.5 Example of risk trend over time adapted from risk tolerability/acceptability levels Fig. 13.3.

values. Fig. 13.5 shows an example of such a risk trend over time. The y-axis is colored according to the risk tolerability/acceptability levels used in the barometer.

Risk trends can also be visualized by means of an inner dial on the barometer that displays the moving average of the risk value over a predetermined period of time, for

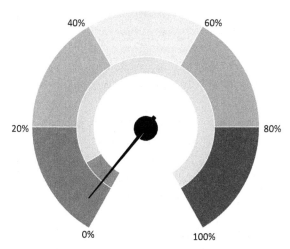

Figure 13.6 Example of risk barometer with moving average over the past 30 days.

example over the past 30 days. In this way, the impression of a decreasing trend such as the one shown in Fig. 13.6 (the moving average over the past 30 days is higher than the current risk level) can be gained at first glance. Moreover, a moving average may mitigate short-term fluctuations and highlight trends in the longer term.

3.1.3 Area Risk Comparison

Chemical and process facilities may be divided into areas with different functions. For instance, an oil platform may be divided into process area, main deck, riser area, central shaft, north shaft, utility area, and so forth.

To compare the risk assessed on a real-time basis for the different areas, a radar chart may be used (Fig. 13.7).

At first glance, Fig. 13.7 shows that most of the risk at the facility is concentrated at areas 1, 2, 3, and 8. The shape and size of the red radar area updates in real time, thereby visualizing the risk variation between the different areas. This figure may provide management and decision-makers with a quick picture of areas that need detailed attention/risk mitigation.

3.2 Drill-down Capability

The risk level visualized through the plots and trends introduced in Section 3.1 is a function of the status and condition of the different barrier functions and associated barrier systems. Each barrier system is modeled and measured through a set of indicators. Drill-down capability enables moving through the hierarchy of the model, from the area of the barrier and further to the indicator level. Information about the overall

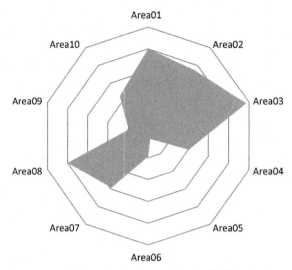

Figure 13.7 Radar chart comparing risk of different areas.

risk, its progression, and underlying causes can be continuously traced, providing intuitive understanding of the causes of risk variations and supporting definition of priorities related to risk mitigation and control.

Fig. 13.8 illustrates the general structure for drill-down functionality in the risk barometer and corresponding visualizations that may be used at each level.

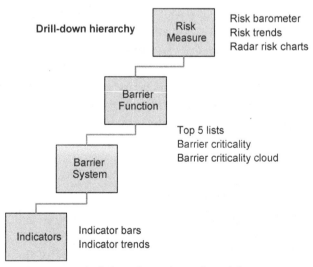

Figure 13.8 Drill-down hierarchy in the risk barometer.

The ultimate manner in which the hierarchy is structured and the types of visualizations used at each level are at the discretion of the end user. The decision context (ie, the nature of decisions to be supported) is the main factor to consider for the structure definition. For instance, facility risk managers would be interested only in the top-level visualizations and historic trends, whereas daily planners would be focusing on the status of barrier functions, systems, and indicator levels.

3.2.1 Top Risk Contributors

To improve decision support in operations, the risk barometer visualizes a list of top risk contributors (Fig. 13.9) using the real-time sensitivity measures described in Section 3.2.2. Such a list highlights which barrier functions and associated barrier systems are contributing the most to the risk level at the given point in time.

3.2.2 Barrier Level

The condition of the different barriers continuously changes on the basis of ongoing activities, deviations, and impairments affecting the indicator values. (The generic term "barriers" is used in this section, but the visualization approach may apply to both barrier functions and barrier systems.) Dynamic sensitivity analysis measures the relative importance of barriers by defining a real-time Birnbaum-like measure based on [8]

$$I^B(i|t) \ = \ \frac{\partial R}{\partial p_i} \ = \ \frac{R\left(p_i^+, t\right) - R\left(p_i^-, t\right)}{p_i^+ - p_i^-} \tag{13.1}$$

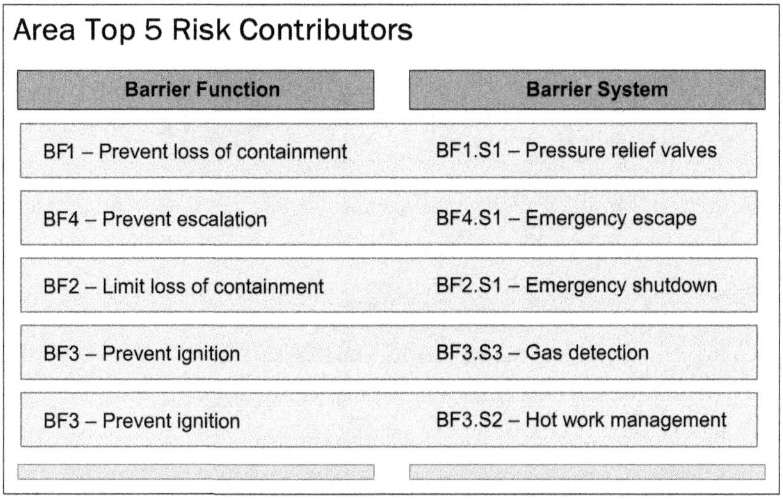

Area Top 5 Risk Contributors	
Barrier Function	**Barrier System**
BF1 – Prevent loss of containment	BF1.S1 – Pressure relief valves
BF4 – Prevent escalation	BF4.S1 – Emergency escape
BF2 – Limit loss of containment	BF2.S1 – Emergency shutdown
BF3 – Prevent ignition	BF3.S3 – Gas detection
BF3 – Prevent ignition	BF3.S2 – Hot work management

Figure 13.9 Top five risk contributors.

where $I^B(i|t)$ = Birnbaum–like measure representing importance of the barrier i at time t; $R(p_i^+, t)$ = conditional risk at time t given the parameter of interest p_i^+ representing the fully degraded barrier i; $R(p_i^-, t)$ = conditional risk at time t given the parameter of interest p_i^- representing the fully operational barrier i; and $p_i^{\pm}$ = parameter of interest representing the barrier i, which may coincide with the barrier failure probability (FP). Fully degraded p_i^+ and fully operational conditions p_i^- can be described by means of the highest and the lowest FPs, respectively, in the realistic variation ranges assessed while building the risk barometer model (see Chapter 7).

However, the Birnbaum–like measure $I^B(i|t)$ reflects the potential of the barrier i to reduce risk at a given time t, but it does not show the status $p_i(t)$ of the barrier i at that given time, which may be represented by its FP. The two pieces of information are equally important to support decision-making. For this reason, the risk barometer allows definition of a two-dimensional diagram plotting real-time evaluation of relative importance (Birnbaum-like measure) and status of barriers (Figs. 13.10 and 13.11).

Traffic light colors are used in the diagram in Figs. 13.10 and 13.11 to visualize in an intuitive way the assessment of safety barriers. Red (dark gray in print versions) indicates an area of high criticality, where relatively important barriers are severely degraded (eg, BF1). Similarly, green (gray in print versions) indicates an area of low criticality, where relatively less important barriers are at optimal status (eg, BF3). Yellow (light gray in print versions) indicates an area of intermediate criticality warning against potential negative developments (eg, BF2 and BF4).

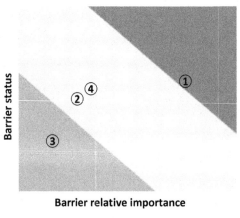

Figure 13.10 Real-time criticality diagram showing relative importance (Birnbaum-like measure) and status of barriers. *LoC,* loss of containment.

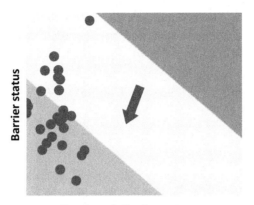

Barrier relative importance

Figure 13.11 Barrier criticality cloud diagram with a trend arrow obtained for "BF3 – prevent ignition".

Barrier criticality may also be expressed through a scatter plot. This allows definition of a "cloud" of the criticality evaluation results collected during a specific time period. An arrow in the middle of the figure indicates whether there has been worsening or improvement in this time period. Fig 13.11 illustrates the improving condition of BF3, the status of which has improved and the importance of which has marginally decreased in the last period.

3.2.3 Indicator Level

Indicators represent the lowest level of information collected in the risk model. Indicator values are translated into mutually comparable scores from 1 to 6 and visualized by means of bars and trend diagrams (Fig. 13.12). Similar to the inner dial on the risk barometer (Fig. 13.6), the triangle on the indicator bars shows the moving average of the indicator value. Moreover, the bars use solid coloring (no traffic light coloring) because the individual indicators are not themselves critical; what is critical is their collective contribution to the risk level.

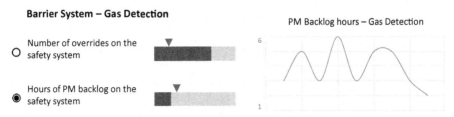

Figure 13.12 Indicator bars and trend diagram. *PM*, preventive maintenance.

Finally, it is important for each indicator to be easily linked and traceable to underlying data sources that supply the indicator measurements. For instance, the indicator "number of overrides on the safety system" in Fig. 13.12 should connect to the list of overrides reported by the control system, and "hours of preventive maintenance backlog" should connect to the list of open preventive maintenance work orders reported by the maintenance management software.

4. CONCLUSIONS

This chapter presents a series of solutions for dynamic risk visualization, which has the potential to overcome common limitations of traditional risk metrics. Such solutions are adopted by the risk barometer approach and include use of the barometer for instantaneous risk visualization, plotting the risk trend over time, listing top risk contributors, creating barrier criticality diagrams, and showing indicator status bars. Clear hierarchy ordinates all the elements and allows the potential user to browse among the risk analysis results and drill down to the aggregated details.

In fact, the ultimate purpose for such visualization solutions is improving the support for critical decision-makers, from risk managers to daily planners. The risk barometer aims not only to offer graphical user interfaces for risk communication but also to continuously update the risk picture on a real-time basis and provide detailed information about the subsystems involved.

REFERENCES

[1] Johansen IL, Rausand M. Foundations and choice of risk metrics. Safety Science 2014;62:386—99.
[2] Delatour G, Laclémence P, Calcei D, Mazri C. Safety performance indicators: a questioning diversity. Chemical Engineering Transactions 2014:36.
[3] Haugen S, Seljelid J, Nyheim OM, Sklet S, Jahnsen E. A generic method for identifying major accident risk indicators. In: 11th international probabilistic safety assessment and management conference and the annual European safety and reliability conference: Curran Associates Inc.; 2012. p. 5743—52.
[4] HSE - Health and Safety Executive, HSL - Health and Safety Laboratory. Societal risk: initial briefing to Societal Risk Technical Advisory Group. Bootle (United Kingdom): HSE; 2009.
[5] TNO. Guidelines for quantitative risk assessment: Purple Book. den Hague (The Netherlands): National Institute of Public Health and Environment (RIVM); 2005.
[6] National Research C. Understanding risk: informing decisions in a democratic society. Washington (DC): The National Academies Press; 1996.
[7] Johansen IL, Rausand M. Risk metrics: interpretation and choice. In: Industrial Engineering and Engineering Management (IEEM), IEEE international conference on 2012; 2012. p. 1914—8.
[8] Rausand M. Risk assessment: theory, methods, and applications. Wiley; 2013.
[9] Franks A. Revised land use planning arrangements around large scale petroleum depots. RR511. Bootle (UK): Health and Safety Executive; 2007.
[10] Jonkman SN, van Gelder PHAJM, Vrijling JK. An overview of quantitative risk measures for loss of life and economic damage. Journal of Hazardous Materials 2003;99:1—30.
[11] Hirst IL, Carter DA. A "worst case" methodology for obtaining a rough but rapid indication of the societal risk from a major accident hazard installation. Journal of Hazardous Materials 2002;92: 223—37.

[12] Boardman AE. Cost-benefit analysis: concepts and practice. Pearson/Prentice Hall; 2006.
[13] Bohnenblust H, Slovic P. Integrating technical analysis and public values in risk-based decision making. Reliability Engineering & System Safety 1998;59:151−9.
[14] Buncefield Major Incident Investigation Board. The Buncefield incident 11 December 2005: the final report of the major incident investigation board. Bootle (UK): HSE Books; 2008.
[15] Health and Safety Executive. Societal risk−technical policy issues. Bootle (UK): HSE; 2010.
[16] Paltrinieri N, Dechy N, Salzano E, Wardman M, Cozzani V. Lessons learned from Toulouse and Buncefield disasters: from risk analysis failures to the identification of atypical scenarios through a better knowledge management. Risk Analysis 2012;32:1404−19.
[17] Falck A, Flage R, Aven T. Risk assessment of oil and gas facilities during operational phase. In: Safety and reliability of complex engineered systems−Proceedings of the 25th European safety and reliability conference, ESREL; 2015. p. 373−80.
[18] Health and Safety Executive. Buncefield: Why did it happen?. Bootle (UK): HSE; 2011.
[19] Statoil. Annual report reporting compendium. Annual report 2012. Oslo, Norway: Statoil; 2013.
[20] Etterlid DE. iSee − visualization of risk related factors. In: IO conference, Trondheim, Norway; 2013.
[21] Hauge S, Okstad E, Paltrinieri N, Edwin N, Vatn J, Bodsberg L. Handbook for monitoring of barrier status and associated risk in the operational phase. SINTEF F27045. Trondheim, Norway: Center for Integrated Operations in the Petroleum Industry; 2015.
[22] Visschers VHM, Meertens RM, Passchier WWF, De Vries NNK. Probability information in risk communication: a review of the research literature. Risk Analysis 2009;29:267−87.
[23] Paltrinieri N, Bucelli M, Grøtan TO, Landucci G. A case of dynamic risk management in the subarctic region. Glasgow: ESREL; 2016.

PART 3

Interaction With Parallel Disciplines

Humans

CHAPTER 14

Human Reliability Analysis: from the Nuclear to the Petroleum Sector

S. Massaiu[1], N. Paltrinieri[2,3]

[1]Institute for Energy Technology, OECD Halden Reactor Project, Halden, Norway; [2]Norwegian University of Science and Technology (NTNU), Trondheim, Norway; [3]SINTEF Technology and Society, Trondheim, Norway

1. INTRODUCTION

A short literature review on human reliability analysis (HRA) methods is provided. Such techniques were initially developed for the nuclear sector and today are also applied to other industrial sectors (eg, oil and gas). It is important to understand what complementary approaches might be available for special applications, such as the different types of scenarios/tasks assessed in the petroleum industry. There are dozens of HRA methods available and published, and finding order in a complex and varied landscape of HRA methods, underlying models, and intended and actual applications is not an easy task. This chapter does not aim to provide a comprehensive review but to help explain the basic differences among methods for the most fundamental HRA questions: How the human contribution to accidents is conceived, the conceptual and empirical bases of quantification, and the process associated with the method's application.

2. BASIC CONCEPTS

HRA is the discipline that provides methods and tools for analyzing and assessing risks caused by operator's actions on a technical system. Fully developed HRA methods date back to the 1970s, when systematic tools for analysis of the operator's contribution to risk were applied in the nuclear industry. There are now several HRA methods available for the nuclear sector, with some being developed for or adapted to other industries as well, such as oil and gas, chemical, and aviation.

2.1 The Probabilistic Risk Assessment Framework

The origin of HRA is in probabilistic risk assessment (PRA), the discipline developed for understanding and quantifying the risks of a serious accident at a nuclear power plant. (The pioneering work is the WASH-1400 report [1].) HRA is the approach used to identify potential operator failures and to systematically estimate the probability of these failures using data, models, or expert judgment.

Dynamic Risk Analysis in the Chemical and Petroleum Industry
ISBN 978-0-12-803765-2

PRA identifies events that may initiate an accident (eg, loss of core coolant, earthquake) and produces sequence models (ie, various sequences of initiating and aggravating events and their outcomes). Sequence models are usually represented as event trees, for which probabilities are given for the likelihood of each event. When an event is a human action or an interaction between humans and machines, appropriate use of HRA provides the necessary input to the overall PRA sequence model: the probabilities for human failure events (HFEs).

In PRA, the human—machine system is analyzed in terms of the interactions between hardware elements and human operators. In the case of the technical element, failures are described in terms of components (eg, pumps and valves) that fail to perform the designed function (ie, failure as malfunction). Similarly, and broadly speaking, humans fail when the tasks assigned are not performed. In traditional HRA the operators are treated as components that perform different functions in different situations (typically functions that could not be automatized at the time the plants were designed). Swain and Guttman [2] noted in the description of the technique for human error rate prediction (THERP), the most influential and still one of the most-used HRA techniques, that:

> The THERP approach uses conventional reliability technology modified to account for greater variability and interdependence of human performance as compared with that of equipment performance [...]. The procedures of THERP are similar to those employed in conventional reliability analysis, except that human task activities are substituted for equipment outputs.

The "operator outputs" (the manifestation of the errors) and the failure mechanisms (the causes for the errors) were recognized as more complex for humans than for technical components. For this reason, a detailed analysis of human failure events was assigned to HRA.

2.2 Human Failure Probabilities: Data and Judgment

HRA started as a purely quantitative discipline to provide the inputs for the PRA, the likelihood of the HFEs represented in the PRA's event trees. The basic problem for HRA was to find the probability that the tasks assigned to the operators would be performed erroneously. The concept of human error probability (HEP) was defined as follows:

$$HEP = \text{Number of times an error has occurred/Number of opportunities}$$
$$\text{for an error to occur}$$

Ideally, such human error rates would be pure actuarial data, consisting of considerable amounts of quantitative data on the performance by nuclear power plant personnel of all tasks relevant to PRA. As a consequence, numerous studies [3—7] were performed to produce data sets or databases to be used as a basis for determining HEPs.

However, a problem of generalizability was understood by the authors of THERP (Swain and Guttmann) [2]. They recognized that HRA analysts had to choose among the HEPs provided, and extrapolation was meaningful only if based on common accepted models of human performance. They also recognized that HRA could not be completely data driven and that expert judgment was unavoidable.

2.3 Human Reliability Analysis Generations

2.3.1 First Generation

A quantitative and data-driven approach to first-generation HRA was chosen for application in PRA (eg, THERP [2]). Simple, often psychologically simplistic, human reliability models were adopted for this purpose: human performance calculated as a linear function of the HEPs for specific actions and performance shaping factors (PSFs) present at the time of action. However, HEPs were typically given before considering the influence of the PSFs, which were more properly adjustment factors. Strictly speaking, the PSFs did not shape performance in the models but rather modified human error probabilities for actions that existed independently of these factors. HEPs for specific actions (or classes of actions) were typically provided in tables separate from the PSF tables [2].

2.3.2 Second Generation

In the 1990s, a reexamination of HRA took place following an increasing number of real PRA applications. Dougherty [8] summarized the concerns about existing HRA methods and practices and urged the development of a second generation of HRA models to overcome the issues. In Dougherty's terms, second-generation HRA models would satisfy the following conditions:

1. *be capable of treating the context of performance and the rich spectrum of human factors in nonsimplistic models*
2. *treat errors of commission*, that is, intentional actions that unwillingly place the system in a situation of higher risk
3. *include latent factors*, that is, errors that are committed before the event and whose effects are not realized until the event occurs
4. *have models of human performance validated beyond statistical pragmatism*, that is, base the method on a human performance model that represents more realistic accident operation

2.4 Human Reliability Analysis Process and Practice

HRA refers to more than just the quantification activity (Fig. 14.1). A typical HRA starts with the definition of a scenario, continues with collection and analysis of qualitative data to identify the key elements for quantification, and concludes with the modeling of all the identified important actions and context conditions (eg, by using fault trees or event

Figure 14.1 Typical human reliability analysis process.

trees to check the logical representation of the scenarios). After quantification, the documented results may be used to update a risk assessment model with HEPs and include development of safety improvement measures.

However, through the years, the HRA community has concentrated mainly on how to obtain HEPs, while overlooking how the overall process was implemented [9]. This lack of recognition has led to continuous development of new HRA methods, addressing only a subset of the process elements, often only quantification, to the point that the term "HRA method" is often understood as the quantification stage alone.

Moreover, HRA is now used in many different areas and fulfills different final purposes. Most HRAs were initially "postinitiator HRAs" focusing on mitigation actions performed in a control room against predefined unwanted events. The spectrum of HRA has now expanded and addresses concerns ranging from the specific risk of fires in nuclear plants (fire HRA [10]) to the overall risk faced on oil and gas production platforms [11]. These new developments require extra requirements for carefully assuring the quality of the HRA process as a whole.

3. REVIEW OF SELECTED HUMAN RELIABILITY ANALYSIS METHODS

This section reviews extensively used methods covering both first- and second-generation HRAs. Recent and comprehensive reviews of these methods can be found elsewhere [12–14].

3.1 Technique for Human Error Rate Prediction

THERP is a comprehensive method for identifying, modeling, and quantifying HFEs in a PRA representing the foundations of HRA methods [2].

3.1.1 Human Error Probabilities

The THERP database of nominal HEPs is one of the primary sources for HRA quantification, independently of the method used. The analysts can use THERP HEPs directly by consulting the tables provided, or indirectly by means of other methods, calculators or software programs. THERP HEPs are the result of expert judgment based on research and practice in human–machine interactions in industrial and military facilities in the 1960s and 1970s. The empirical basis is sparse with data that is not easily traceable [13].

3.1.2 Process and Practice

THERP is resource intensive and intended for use by human factor experts. It is probably the most-used HRA technique worldwide for nuclear applications. It is also used in the offshore chemical process and other industries. Despite being one of the oldest methods (with limitations regarding cognitive aspects of operator performance, errors of commission, and the context of performance), it has a high level of acceptance within the international HRA community.

3.2 Human Error Assessment and Reduction Technique

The human error assessment and reduction technique (HEART) aims to be a quick and simple quantification method applicable to any situation or industry where human reliability is important [15,16].

3.2.1 Human Error Probabilities

This method provides nominal HEPs for generic task types for which the empirical basis is expert judgment. Nominal conditions represent the cases in which reliability tends to be achieved consistently. The human reliability predicted for a given action needs to be adjusted by the actual conditions for the task. This is obtained by evaluating the degree of presence of predefined error-producing conditions (the HEART name for PSFs).

3.2.2 Process and Practice

HEART has been extensively used in the UK, originally in the nuclear industry and later in most other industries (eg, chemical, aviation, rail, medical). It is a quantification method that does not require a detailed decomposition of the tasks because it quantifies at the level of generic task types. For this reason it could lend itself to "quick and dirty" applications. However, the UK HRA community typically applies it in a broader assessment framework (eg, safety cases), placing a strong emphasis on performing a thorough task analysis prior to quantification (including plant visits, interviews with plant experts, and performance observations) and on documenting it [12].

3.3 A Technique for Human Event Analysis

A technique for human event analysis (ATHEANA) is a method developed in the mid-1990s that meets Dougherty's four criteria [8] for qualifying as a second-generation technique [17].

3.3.1 Human Error Probabilities

ATHEANA does not provide nominal or basic HEPs for operator tasks. The probability of HFEs is quantified based on (1) estimates of how likely or how frequently plant conditions and other contextual factors affecting HFEs occur, and (2) how likely the errors will occur in the given circumstances. Expert elicitation is suggested in the form of a facilitator-led group consensus procedure. Guidance is given for defining the scale to be used in the uncertainty distribution, but the actual combination of event contributors into a single distribution is left to the analysts' discretion.

3.3.2 Process and Practice

Few real-case applications were made by means of ATHEANA [12]. Its development was commissioned at the same time as the development of the standardized plant analysis risk (SPAR) models by the Idaho National Laboratory, which produced the HRA tool SPAR-H [18]. As a quick and purely quantitative method, SPAR-H gained more diffusion than ATHEANA. However, ATHEANA's popularity has recently increased together with the wider use of probabilistic risk approaches for various type of analysis. For instance, it was used for fire risk analysis (fire HRA [10]) and for integration of purely quantitative methods (eg, the SPAR-H manual recommends ATHEANA [18]).

3.4 Méthode d'Evaluation de la Réalisation des Missions Opérateur pour la Sûreté

Nowadays there are different versions of méthode d'evaluation de la réalisation des missions opérateur pour la sûreté (MERMOS; in English: assessment method for the performance of safety-related operator missions) for detailed, simplified, and statistical analysis as well as for different types of application areas. This description is based on the main module [19].

3.4.1 Human Error Probabilities

MERMOS does not provide nominal or basic HEPs for operator tasks. The probability of an HFE (defined as "mission failure" in MERMOS) is the sum of the probabilities of the (mutually exclusive) failure scenarios described for the HFE (ie, different ways of failing to achieve the mission). These in turn are calculated from the probabilities of occurrence of the situational features and crew operational modes in the scenario. The probabilities of timely recovery (reconfiguration) failure and nonidentifiable scenarios (residual probability) are also estimated. Input probabilities are based on statistical data (if available) or analyst judgment.

3.4.2 Process and Practice

The MERMOS process is described as consisting of five overall stages: HFE identification, design of the scenarios that led to HFE failure, verification of the completeness and coherence of the scenario sets, quantification, and final verification/adjustment. It is a nuclear-specific tool and has been applied only by the Électricité de France staff. In fact, it is a proprietary tool, and not all documentation is available in English.

On the positive side, it is based on a generic model of at-risk systems that can be adapted to other industries [20].

4. HUMAN RELIABILITY ANALYSIS APPLICATION IN THE PETROLEUM SECTOR

Human factors have been shown to play an important role in both the cause and the mitigation of petroleum industry major accidents, as demonstrated by the Deepwater Horizon (Gulf of Mexico) accident [21]. However, the focus of quantitative risk assessment for this industrial field has been traditionally on technical systems and capabilities.

For this reason, the attention dedicated to human and organizational factors in oil and gas industries is gradually increasing, following several related recommendations suggesting their inclusion [22]. For instance, as part of Statoil (Norway) overall safety strategy, human reliability assessments are being applied in major accident risk analyses of offshore activities, such as drilling and production of oil [23]. The quantitative technique SPAR-H is currently being used in the petroleum industry context, and a new methodology denominated Petro-HRA (HRA for the petroleum industry) is under development.

4.1 Standardized Plant Analysis Risk—Human Reliability Analysis Method

The standardized plant analysis risk—human reliability analysis (SPAR-H) method is a quantitative method that is popular in both its intended area of use (US nuclear power plants) and other industries [18]. To estimate HEPs, a nominal error probability is multiplied by eight PSFs representing aspects of individual characteristics, environment, organization, or tasks decrementing or improving human performance. SPAR-H was created as a simplified version of THERP. It has been criticized for relying excessively on judgment from analysts as a result of lack of guidance on the PSF assessment and the remaining qualitative steps of the HRA process [24]. One of the strengths of SPAR-H is simplicity, leading to clear indications of the potential human issues affecting the overall risk [24]. A detailed description of SPAR-H application steps is provided in the following chapter.

4.2 Petro—Human Reliability Analysis

A project funded by The Research Council of Norway is evaluating and adjusting the steps of HRA to suit the needs of the petroleum industry. The outcome of the project will be a guideline for a new HRA method called Petro-HRA [25]. This method will include guidelines on how to collect data, model an event, and quantify the related HEPs by means of a nominal HEP and nine PSFs. Quantification is based on the SPAR-H technique and the experience developed through its application in this sector.

5. CONCLUSIONS

HRA started as a purely quantitative discipline that strictly followed the PRA's footsteps. The HRA community was aware of the simplifying assumptions made when treating humans as technical components, but the empirical validity of HRA estimates began to be questioned only in the 1990s. This was due to the accumulated experience with and range of HRA uses, which had considerably expanded. A new generation of methods, developed for deeper integration with risk analyses, could better model complexity of human performance. Today HRA is used in a variety of different application areas, from the nuclear to the petroleum sector. The fundamental importance of human factors in the prevention of oil and gas major accidents is recognized not only by the research community but also by oil companies: the former by defining *ad hoc* projects for the adjustment of HRA techniques originally developed for the nuclear sector, and the latter by including their application in their overall safety strategy.

REFERENCES

[1] U.S. Nuclear Regulatory Commission. Reactor safety study: an assessment of accident risks in U.S. commercial nuclear power plants. WASH-1400, NUREG-75/014. Washington, DC, U.S.: NRC; 1975.

[2] Swain AD, Guttman HE. Handbook of human reliability analysis with emphasis on nuclear power plant applications. Sandia National Labs NUREG CR-1278. Washington, DC, U.S.: Nuclear Regulatory Commission; 1983.

[3] Beare AN, Dorris RE, Bovell CR, Crowe DS, Kozinsky EJ. A simulator-based study of human errors in nuclear power plant control room tasks (NUREG/CR-3309). Washington, DC, U.S.: U. S. Nuclear Regulatory Commission; 1984.

[4] Beare A, Gaddy C, Singh A, Parry GW. An approach for assessment of the reliability of cognitive response for nuclear power plant operating crews. In: Apostolakis G, editor. Proceedings of probabilistic safety assessment and management. Beverly Hills, CA, U.S.: Elsevier; 1992.

[5] Ghertman F, Griffon-Fouco M. Investigation of human performance events at French power stations. In: IEEE third conference on human factors and nuclear safety, Montery, CA, U.S.; 1995.

[6] Weston LM, Whitehead DW, Greaves NL. Recovery actions in PRA for the risk methods, integration and evaluation program (RMIEP). In: Development of the data base method. NUREG/CR-4834, vol. I. Washington, D.C., U.S.: US Nuclear Regularity Commission; 1987.

[7] Moroni JM, Mosneron-Dupin F, Villemeur A, Meslin T. Simulator-based study of nuclear plants operators' behaviour under abnormal conditions. Bulletin de la Direction des Etudes et Recherches Serie A, Nucleaire, Hydraulique, Thermique 1987;1:69–73.

[8] Dougherty EM. Human reliability analysis—where shouldst thou turn? Reliability Engineering and System Safety 1990;29:283–99.

[9] Forester J, Kolaczkowski A, Lois E, Kelly D. Evaluation of human reliability analysis methods against good practices. No. NUREG-1842. Washington DC, U.S.: U.S. Nuclear Regulatory Commission; 2006.

[10] Lewis S, Cooper S. EPRI/NRC-RES fire human reliability analysis guidelines − final report (NUREG-1921). Washington DC, U.S.: U.S. Nuclear Regulatory Commission; 2012.

[11] Laumann K, Rasmussen M. Suggested improvements to the definitions of standardized plant analysis of risk-human reliability analysis (SPAR-H) performance shaping factors, their levels and multipliers and the nominal tasks. Reliability Engineering and System Safety 2016;145:287–300.

[12] Bell JB, Holroyd J. Review of human reliability assessment methods. vol. RR679. Bootle, UK: Health and Safety Executive; 2009.

[13] Forester J, Dang VN, Bye A, Lois E, Massaiu S, Broberg H, et al. The International HRA Empirical Study — final report — lessons learned from comparing HRA methods predictions to HAMMLAB simulator data. HPR-373, NUREG-2127. Halden, Norway: OECD Halden Reactor Project; 2013.

[14] Chandler FT, Chang YHJ, Mosleh A, Marble JL, Boring RL, Gertman DI. Human reliability analysis selection guidance for NASA. Washington DC, U.S.: National Aeronautics and Space Administration; 2006.

[15] Williams JCA. Data-based method for assessing and reducing human error to improve operational experience. In: IEEE 4th conference on human factors in power plants. Monterey, California, U.S.: IEEE; 1988.

[16] Williams JC. A proposed method for assessing and reducing human error. 9th Advance in Reliability Technology Symposium. Bradford, UK: University of Bradford; 1986.

[17] Cooper SE, Ramey-Smith AM, Wreathall J, Parry GW, Bley DC, Luckas WJ, et al. A technique for human event analysis (ATHEANA) — technical basis and methodological description. NUREG/CR-6350. Upton, New York, U.S.: U.S. Nuclear Regulatory Commission, Brookhaven National Laboratory; 1996.

[18] Gertman D, Blackman H, Marble J, Byers J, Smith C. The SPAR-H human reliability analysis method: Division of Risk Analysis and Applications Office of Nuclear Regulatory Research U.S. Nuclear Regulatory Commission. 2005.

[19] Le Bot P, Desmares E, Bieder C, Cara F, Bonnet J-L. MERMOS: an EdF project to update the PHRA (probabilistic human reliability assessment) methodology. Chattanooga, Tennessee, U.S.: OECD Nuclear Energy Agency Specialists Meeting on Human Performance in Operational Events; 1997.

[20] Le Bot P. Analysis of the Scottish case. In: Hollnagel E, Nemeth CP, Dekker S, editors. Resilience engineering perspectives: remaining sensitive to the possibility of failure. Farnham, UK: Ashgate Publishing, Ltd; 2008.

[21] Norazahar N, Khan F, Veitch B, MacKinnon S. Human and organizational factors assessment of the evacuation operation of BP Deepwater Horizon accident. Safety Science 2014;70:41—9.

[22] Skogdalen JE, Vinnem JE. Quantitative risk analysis offshore—human and organizational factors. Reliability Engineering and System Safety 2011;96:468—79.

[23] Gould KS, Ringstad AJ, Van De Merwe K. Human reliability analysis in major accident risk analyses in the Norwegian petroleum industry. In: Proceedings of the human factors and ergonomics society; 2012. p. 2016—20.

[24] Forester J, Dang V, Bye A, Boring R, Liao H, Lois E. Conclusions on human reliability analysis (HRA) methods from the International HRA Empirical Study. In: ESREL 2012 and Probabilistic Safety Assessment and Management Conference (PSAM) 11, Helsinki, Finland; 2012.

[25] Laumann K, Øien K, Taylor C, Boring RL, Rasmussen M. Analysis of human actions as barriers in major accidents in the petroleum industry, applicability of human reliability analysis methods (Petro-HRA). PSAM 2014-Probabilistic Safety Assessment and Management; 2014.

CHAPTER 15

Human Reliability Analysis in the Petroleum Industry: Tutorial and Examples

N. Paltrinieri[1,2], S. Massaiu[3], A. Matteini[4]

[1]Norwegian University of Science and Technology (NTNU), Trondheim, Norway; [2]SINTEF Technology and Society, Trondheim, Norway; [3]Institute for Energy Technology, OECD Halden Reactor Project, Halden, Norway; [4]University of Bologna, Bologna, Italy

1. INTRODUCTION

Quantitative risk assessment (QRA) is a consolidated approach to evaluating the risk level of an industrial system, which is traditionally based on the main technical failures leading to potential accident scenarios. However, such scenarios may be the result of interaction among a series of elements, which range from the technical to the human and organizational domains. Human and organizational factors often have an important role in the development of a scenario [1], and their assessment is essential for accurate QRA and effective risk mitigation. For this reason, this chapter addresses the integration of QRA by means of human reliability analysis (HRA), which not only assesses the probability of the identified human failure events (HFEs, or failures of functions, systems, or components that are the result of human errors) but also evaluates the influence of underlying factors on human performance.

2. METHODOLOGY TUTORIAL

HRA typically includes definition of a specific scenario to analyze (which may be suggested by the QRA of the system studied), collection of relevant qualitative data, study of human actions (tasks) involved, and potential related human errors, which are then modeled and quantified to outline tailored measures of human error reduction (Fig. 15.1). The phase of quantification produces an estimate of the probability of one or several human errors (HFEs), which may integrate the QRA phase of analysis of initiating events and their frequencies (see Fig. 15.1), as indicated by the NORSOK standard on "Risk and emergency preparedness analysis" [2]. A specific methodology for HFE quantification prevalently used in the nuclear sector is the standardized plant analysis risk—human reliability analysis (SPAR-H) [3]. Owing to its relative simplicity, this method is progressively being used for HRA in the oil and gas sector [4,5].

Dynamic Risk Analysis in the Chemical and Petroleum Industry
ISBN 978-0-12-803765-2

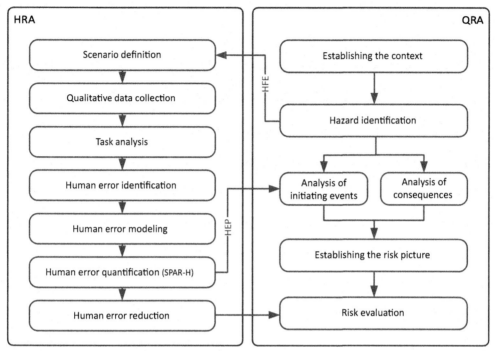

Figure 15.1 Mutual integration of human reliability analysis and quantitative risk assessment.

This chapter aims to describe SPAR-H and give an example of its application in the oil and gas sector, but it does not provide guidance on how to identify or model HFEs. HFEs may be suggested by the phase of hazard identification of the related QRA (see Fig. 15.1) and further modeled in detail in the first phases of an HRA (as shown in Section 3). Chapter 14 reports a brief overview of HRA methods that may be consulted for such activity.

For this reason, the following tutorial describes the application of SPAR-H for the quantification of a generic human error, which is assumed to be adequately identified and modeled (Fig. 15.2).

2.1 Step 1: Human Error Categorization

SPAR-H categorizes human errors as either diagnosis tasks or action tasks (or combined diagnosis and action). Diagnosis refers to cognitive processing, from interpreting information to deciding to act. Action refers to simple action implementation, such as pressing a button or turning a dial. Whaley et al. [3] suggests that there are very few situations where a diagnosis and an action are not linked. However, despite the fact that action rarely occurs without diagnosis, it might be possible to have a diagnosis that is not followed by an action.

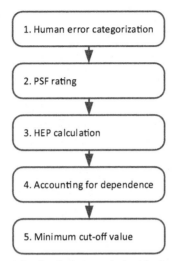

Figure 15.2 Steps in standardized plant analysis risk—human reliability analysis.

2.2 Step 2: Performance-Shaping Factor Rating

The context associated with human errors (from the technical to the organizational underlying factors) is taken into account by using the eight performance-shaping factors (PSFs) suggested by SPAR-H. The following questions should be answered first for each PSF:

- Is there adequate information to judge the influence of the PSF?
- Does the PSF exert a significant influence on the likelihood of failure for the human operator?

If there is a negative answer to one of the questions, the PSF should not be considered (by either assigning nominal value or excluding it). If both the answers are affirmative, the PSF should be evaluated by choosing the most appropriate level for the context. Description of PSFs and related levels are reported in Table 15.1.

2.3 Step 3: Human Error Probability Calculation

The human error probability (HEP) studied is calculated as follows:

$$HEP = NHEP \cdot \prod_{i=1}^{8} PSFM_i \qquad (15.1)$$

$$NHEP = 1.0 \cdot 10^{-2} \text{ for diagnosis}$$

$$NHEP = 1.0 \cdot 10^{-3} \text{ for action}$$

where $NHEP$ is the nominal HEP, differing between diagnosis and action, and $PSFM_i$ is the multiplier of ith PSF. Although other formulas may be used to avoid results higher than unity [3,6], it is common practice to approximate such results to a probability of failure equal to 1.

Table 15.1 Performance Shaping Factors in Standardized Plant Analysis Risk–Human Reliability Analysis [6]

PSF	Description	Diagnosis Levels	Action Levels	Multipliers
1. Available time	Amount of time t that an operator or a crew has to diagnose and act on an abnormal event. A shortage of time can affect the ability to think clearly and consider alternatives. It may also affect the ability to perform an action.	Inadequate time $t \sim 2/3T$ Nominal time T $T < t < 2T$ and $t > 30$ min $t > 2T$ and $t > 30$ min	Inadequate time $t \sim T$ Nominal time T $t \geq 5T$ $t \geq 50T$	P (failure) = 1 10 1 0.1 0.01
2. Stress/ stressors	Level of undesirable conditions and circumstances (mental or physical) that impede the operator from easily completing a task.	Extreme High Nominal	Extreme High Nominal	5 2 1
3. Complexity	How difficult the task is to perform in the given context. It considers both the task and the environment in which it is to be performed. Ambiguity, mental elaborations, and physical efforts are considered.	Highly complex Moderately complex Nominal Obvious diagnosis	Highly complex Moderately complex Nominal /	5 2 1 0.1
4. Experience/ training	Years of experience of the individual or crew, whether they have been trained for the type of accident, the amount of time passed since training, and the systems involved in the task and scenario.	Low (diagnosis) / Nominal High	/ Low (action) Nominal High	10 3 1 0.5
5. Procedures	Existence and use of formal operating procedures for the tasks under consideration. Wrong or inadequate information and ambiguity of steps are considered.	Not available Incomplete Available, but poor Nominal Diagnostic oriented	Not available Incomplete Available, but poor Nominal /	50 20 5 1 0.5

Table 15.1 Performance Shaping Factors in Standardized Plant Analysis Risk—Human Reliability Analysis [6]—cont'd

PSF	Description	Diagnosis Levels	Action Levels	Multipliers
6. Ergonomics/ human— machine interface	Equipment displays and controls, layout, and quality and quantity of information available from instrumentation and the interaction of the operator/crew with the equipment to carry out tasks.	Missing/ misleading Poor Nominal Good	Missing/ misleading Poor Nominal Good	50 10 1 0.5
7. Fitness for duty	Whether the individual performing the task is physically and mentally fit to perform the task at the time. Fatigue, sickness, drug use (legal or illegal), overconfidence, personal problems, and distractions are considered.	Unfit Degraded fitness Nominal	Unfit Degraded fitness Nominal	P (failure) = 1 5 1
8. Work processes	Aspects of doing work, including interorganizational safety culture, work planning, communication, and management support and policies. How work is planned, communicated, and executed is considered.	/ Poor (diagnosis) Nominal Good (diagnosis) /	Poor (action) / Nominal / Good (action)	5 2 1 0.8 0.5

2.4 Step 4: Accounting for Dependence

Dependence exists when the occurrence of one event affects the likelihood of a second event. By definition, HFEs and their respective HEPs should be independent of one another. To determine whether dependence is present between HFEs, the analyst must assess the following:

- the sequence of events that has led to this point in the accident scenario;
- important plant/equipment status and performance;
- the context surrounding the tasks described in the HFE (related performance drivers and causal connections from previous activities and/or equipment issues).

However, an HFE may be composed of a series of events to which it is obviously dependent. This may be accounted for by means of fault tree logic modeling, as shown in Fig. 15.5. To calculate the overall HEP, event probabilities are aggregated by means of appropriate logic gates ("AND" and "OR" gates) and related Boolean algebra [7], as follows:

- OR gate − logical inclusive disjunction, $P(A \text{ or } B) = P(A) + P(B)$
- AND gate − logical conjunction, $P(A \text{ and } B) = P(A) \cdot P(B)$

2.5 Step 5: Minimum Cut-off Value

If the calculated HEPs are lower than about 10^{-6}, the importance of related failure mechanisms that would otherwise be judged to be negligible (eg, heart attack of the operator) sensibly increases. However, there might a virtually infinite number of potential failure mechanisms. The smaller the probabilities, the more error modes should be considered. For this reason, the lower bound on a single HEP in SPAR-H is suggested to be 10^{-5}. The calculated HEPs falling under this limit should be disregarded (or reassessed).

3. APPLICATION OF SPAR-H

3.1 Drive-off of a Semisubmersible Drilling Unit

This application example addresses a scenario of drive-off of a semisubmersible drilling unit located in Norwegian shallow waters. To avoid potential damage, the rig should maintain the position above the wellhead, where the drilling operations are conducted. Such positioning is maintained in an autonomous way (without mooring system) through the action of a set of thrusters controlled by the dynamic positioning (DP) system. Operations in shallow waters admit low tolerance in terms of rig position owing to potentially higher riser angles [8]. Such a scenario is relatively more critical in Norwegian waters, where shallow water operations are classified as operations in 320-m-deep waters or less. (Internationally the limit is about 600 m).

Input for the DP system is provided by the position reference system (differential global positioning system and hydroacoustic position reference), environmental sensors, gyrocompass, radar, and inclinometer [8]. A dynamic positioning operator (DPO) located in the marine control room is responsible for constant monitoring of DP panels and screens and carrying out emergency procedures if needed [9].

Platform position may be lost for a series of reasons. In this example it is assumed that the platform thrusters exercise propulsion toward a wrong direction, leading to a scenario of drive-off. To establish whether the rig is located in the safe-operation area, specific offset position limits are drawn (Fig. 15.3). Such limits are defined taking into account

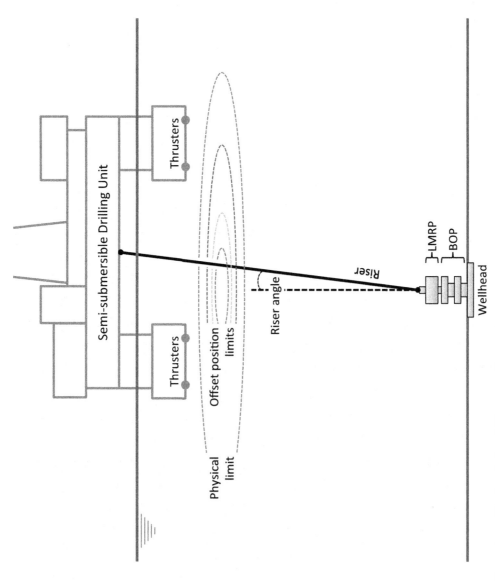

Figure 15.3 Dynamic positioning during drilling operations.

riser angle, position data, and environmental variables. The riser has relatively low capacity of inclination, despite the upper and lower flex joints, and it can reach a maximum angle of 12 degrees. Exceeding this physical limit may result in damages to the wellhead, the blowout preventer (BOP) sealing the well, or the lower marine riser package connecting the riser and the BOP [8]. For this reason, a conservative maximum angle of 8 degrees is considered in the limit calculation.

If the rig moves to an offset position, specific alarms turn on to suggest to the DPO to initiate the manual emergency disconnect sequence (EDS) for the disconnection of the riser from the BOP. If the manual EDS ultimately fails, the automatic EDS activates at the ultimate position limit allowing for safe disconnection (red limit) [8]. Operations in shallow waters imply shorter available time for detection and position recovery for both system and operators. For this reason, automatic EDS has to be always enabled in shallow waters.

3.2 Human Failure Event Identification and Modeling

In a drive-off scenario of a semisubmersible drilling unit, a standard QRA would most likely consider that the technical safety barriers will act to prevent well damage. However, safety barriers are not only physical and technical, but also human and operational, systems aimed to prevent, control, or mitigate undesired events [10]. In fact, the two main tasks to be performed by a DPO in such a scenario are the following:

1. stopping the drive-off condition by pushing the emergency stop buttons of the working thrusters;
2. manually initiating the EDS process by pushing the EDS button.

Such tasks represent two human or operational barriers. In particular, manual EDS is considered the primary barrier by the DPO Committee of the Marine Technology Society [11], whereas automatic EDS is considered a redundant safety barrier. On the basis of these considerations, the event tree represented in Fig. 15.4 was defined to describe the scenario.

For the sake of brevity, this application example focuses only on the manual EDS barrier (barrier 2 in Fig. 15.4—but the same analysis can be also performed on barrier 1), for which the related HFE is defined as "failure to manually initiate EDS." This HFE can be modeled by means of a fault tree diagram to highlight the subevents leading to it (Fig. 15.5). In fact, the DPO may fail to initiate the EDS in time (basic event [BE] 1) or not initiate the EDS at all (BE2), and a series of error modes may be, in turn, their causes. The DPO may either fail to detect the first set of alarms activating in such a scenario (high thruster force alarms—E1.1) or take too long while deciding (E1.2) or acting (E1.3). Each of these error modes would most certainly lead to a vain attempt to initiate

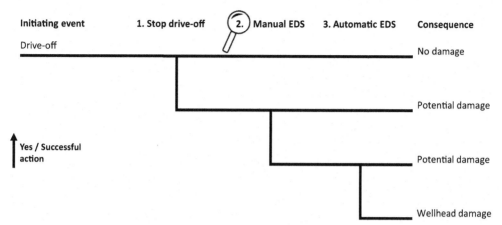

Figure 15.4 Event tree describing a drive-off scenario.

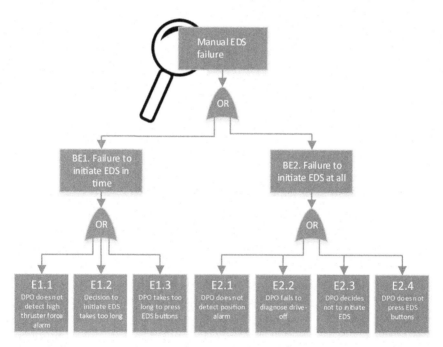

Figure 15.5 Fault tree describing the human failure event of a manual emergency disconnect system failure.

EDS owing to the shortage of time (BE1). The DPO may also either fail to detect the last set of alarms activating in such a scenario (position alarms—E2.1) or fail in the diagnosis (E2.2), decision (E2.3), or action (E2.4). Each of these error modes would lead to the failure to initiate EDS at all (BE2).

3.3 Human Error Probability Quantification

SPAR-H is a technique of HEP quantification that can be applied to quantify the events modeled in the fault tree in Fig. 15.5. Although HEP quantification can be performed for the elements on the lowest fault tree level (error mode E) and the probabilities can be aggregated following the logic gates used, the decision on the level at which to apply SPAR-H is at analyst's discretion. In fact, the analyst may find it more effective to focus directly on a higher level (BE or HFE).

In this example, SPAR-H is applied on the seven error modes reported in the fault tree in Fig. 15.5, and the results are reported in Table 15.2.

The first step is to classify the error modes as either diagnoses or actions. Only two error modes cannot be defined as cognitive processes, but they are clearly the implementation of previous cognitive processes, that is, actions: E1.3 (the DPO takes too long to press the EDS button) and E2.4 (the DPO does not press the EDS button). The other error modes are classified as diagnoses (Table 15.2).

Qualitative information on the scenario (assumed to be collected in the previous HRA phases [see Fig. 15.1]) is used to study and rate the PSFs for each error mode considered, as shown in Table 15.2. The context of E1.1, "DPO does not detect high thruster force alarm," and E2.1, "DPO does not detect position alarm," may be described by the ergonomics/human–machine interface PSF owing to the potential of alarm flooding. However, no information demonstrating such a hypothesis is collected, and the nominal level is selected. The context of E1.2, "Decision to initiate EDS takes too long," and E2.3, "DPO decides not to initiate EDS," may be mainly described by the stress/stressors PSF, because the decision to initiate the EDS led to financial consequences to the company and high stress. The context of E1.3, "DPO takes too long to press EDS button," and E2.4, "DPO does not press EDS button," may be mainly described by the available time PSF owing to the shortage of time to take this action (about the time required). The context of E2.2, "Fails to diagnose drive-off," may be better described by two driving PSFs: experience/time and work processes. The first PSF accounts for the rarity of such a scenario, leading to relatively low experience. The second PSF accounts for the collaboration between the DPOs on the bridge, which would increase the probability of correctly diagnosing drive-off.

Once the PSF levels are selected, the Eq. (1) is applied to assess the probability of the single error modes. The dependence of the BEs on the assessed error modes is accounted for by considering the logic gate linked to them and following the related aggregation rules. In this case only "OR" gates are used, and the probabilities of E1.1–E1.3 and E2.1–E2.4 are summed to obtain the probability of BE1 and BE2, respectively. In turn, the probabilities of BE1 and BE2 are summed (OR gate) to obtain the overall HEP of "Failure to manually initiate EDS." This HEP may be eventually employed for barrier 2 in the event tree analysis represented by Fig. 15.4.

Table 15.2 Human Error Quantification Results

Error mode	Diagnosis/ Action	Performance Shaping Factors	PSF Multiplier	Error Mode Probability	Basic Event	Basic Event Probability	HFE	HEP
E1.1	Diagnosis	Ergonomics/human–machine interface	1	0.01	BE1	0.04	Failure to manually initiate emergency disconnect system	0.16
E1.2	Diagnosis	Stress/stressors	2	0.02				
E1.3	Action	Available time	10	0.01				
E2.1	Diagnosis	Ergonomics/human–machine interface	1	0.01	BE2	0.12		
E2.2	Diagnosis	Experience/time	10	0.08				
		Work processes	0.8					
E2.3	Diagnosis	Stress/stressors	2	0.02				
E2.4	Action	Available time	10	0.01				

4. CONCLUSIONS

This chapter presents a step-by-step tutorial for the application of the SPAR-H technique. SPAR-H may be employed to integrate the QRA of a system by quantifying the human contribution to risk. Underlying factors affecting human performance are taken into account by rating relevant PSFs. However, the application of SPAR-H should be preceded by thorough qualitative analysis of the tasks performed by the operator and crew in the scenario considered. For this reason, coupling with one of the HRA techniques mentioned in the previous chapter is suggested.

An example of application of SPAR-H and integration of a QRA event tree is shown. A scenario of drive-off of a semisubmersible drilling unit is studied to show how to account for issues of human—machine interface, stress, shortage of time, and rare-event experience, which may lead the DPO into failure and increase the probability of an accident.

REFERENCES

[1] Schönbeck M, Rausand M, Rouvroye J. Human and organisational factors in the operational phase of safety instrumented systems: a new approach. Safety Science 2010;48:310—8.
[2] NORSOK. Risk and emergency preparedness assessment. Oslo, Norway: Standards Norway; 2010.
[3] Whaley AM, Kelly DL, Boring RL, Galyean WJ. SPAR-H step-by-step guidance, INL/EXT-10—18533. Rev. 2. Idaho Falls, Idaho, U.S.: Idaho National Laboratory; 2011.
[4] Rasmussen M, Standal MI, Laumann K. Task complexity as a performance shaping factor: a review and recommendations in standardized plant analysis risk-human reliability analysis (SPAR-H) adaption. Safety Science 2015;76:228—38.
[5] Laumann K, Øien K, Taylor C, Boring RL, Rasmussen M. Analysis of human actions as barriers in major accidents in the petroleum industry, applicability of human reliability analysis methods (Petro-HRA). PSAM 2014-Probabilistic Safety Assessment and Management 2014.
[6] Gertman D, Blackman H, Marble J, Byers J, Smith C. The SPAR-H human reliability analysis method. Idaho Falls, Idaho, U.S: Idaho National Laboratory; 2005.
[7] Center for Chemical Process Safety. Guidelines for chemical process quantitative risk analysis. New York, USA: America Institute of Chemical Engineers (AIChE); 2000.
[8] Chen H, Moan T, Verhoeven H. Safety of dynamic positioning operations on mobile offshore drilling units. Reliability Engineering and System Safety 2008;93:1072—90.
[9] Giddings IC. IMO Guidelines for vessels with dynamic positioning systems. In: Dynamic Positioning Committee — Marine Technology Society. Dynamic Positioning Conference, Houston, Texas, U.S.; 2013.
[10] Sklet S. Safety barriers: definition, classification, and performance. Journal of Loss Prevention in the Process Industries 2006;19:494—506.
[11] Dynamic Positioning Committee. DP operations guidance. Washington, DC, U.S.: Marine Technology Society; 2012.

Costs and Benefits

CHAPTER 16

Cost-Benefit Analysis of Safety Measures

G. Reniers[1,2,3], L. Talarico[1], N. Paltrinieri[4,5]

[1]University of Antwerp, Antwerp, Belgium; [2]KU Leuven, Brussels, Belgium; [3]Delft University of Technology, Delft, The Netherlands; [4]Norwegian University of Science and Technology (NTNU), Trondheim, Norway; [5]SINTEF Technology and Society, Trondheim, Norway

1. INTRODUCTION

At first look, it seems evident that consideration of both risk management and safety management is essential in any manager's decision. However, Perrow [1] indicates that there are indeed reasons why managers and decision-makers would not put safety first. A very important reason is that harm and consequences are not evenly distributed: The latency period may be longer than any one decision-maker's career. Few managers are punished for not putting safety first, even after an accident, but will quickly be punished for not putting profits, market share, or prestige first. In the long term, this approach would obviously not be the best management solution for an organization.

A central concept is that of cost. On the one hand, there are the costs of improving work conditions to reduce the incidence of injury and disease. On the other hand, there are the costs resulting from not doing these things. Therefore, a discussion of operational safety within an organization always involves a discussion of choices. One should realize that safety has not only costs but also benefits. Benefits are much harder to acknowledge by managers because they have a hypothetical and uncertain nature. At the end of a working day, employees from a company will not ask themselves, "How many times was I not injured at work today?"

Companies may face many risks but do not often possess adequate information for risk assessment. In particular, this chapter focuses on risks occurring when few or extremely few historical data are available. This kind of risk has the potential to result in catastrophes with major consequences and often with multiple fatalities, so-called low-probability, high-impact events. They do occur on a (semi)regular basis from a worldwide perspective, and large fires, large releases, explosions, toxic clouds, and so forth belong to this class of accidents. For this reason, the associated risk has to be reduced to the lowest practicable level, taking into account the costs and the benefits of such risk reduction. To this extent, different economic considerations should be made, and a specific economic analysis should be carried out.

Dynamic Risk Analysis in the Chemical and Petroleum Industry
ISBN 978-0-12-803765-2

2. THE FOUNDATIONS OF COST-BENEFIT ANALYSIS

In this section a brief overview of the approach to economic evaluation of safety investment is provided. The approach is based on evaluation of costs, benefits of safety measures, and analysis of safety investment.

2.1 Cost-Benefit Analysis Approach

A cost-benefit analysis is an economic evaluation in which all costs and consequences of a certain decision are expressed in the same units, usually money [2,2a]. Such an analysis may be employed in relation to operational safety, to aid normative decisions about safety investments. One should keep in mind that it cannot be demonstrated whether one safety investment is intrinsically better than another. Nevertheless, a cost-benefit analysis allows decision-makers to improve their decisions by adding appropriate information on costs and benefits to certain investment decisions regarding prevention or mitigation. Because decisions about safety investments involve choices between different possible risk management options, cost-benefit analysis can be very useful. Moreover, decisions may not be straightforward in some cases, especially in the process industries, where there are important risks to be managed and controlled. However, it is worth mentioning that cost-benefit analysis is not a pure science and sometimes needs to employ debatable concepts such as the value of human life, the value of body parts, and the question of who pays the prevention costs and who receives the safety benefits.

Two types of cost-benefit analyses are available. First, there is the "*ex ante* cost-benefit analysis," which is carried out before deciding about whether to invest. Second, the "*ex post* cost-benefit analysis" is conducted at the end of a project [3] to verify the profitability of the project. Although in this chapter we are mainly interested in *ex ante* cost-benefit analysis, in which safety managers complete an economic exercise to aid in the decision-making process with respect to safety investments, the technique described can also be used to carry out an *ex post* cost-benefit analysis. In any case, the result of any *ex ante* cost-benefit analysis, being a recommendation for an investment decision regarding prevention, is meant to merely assist the decision-maker in the decision process by making costs and benefits more transparent and more objective.

The decision-maker is recommended to use this approach with caution because the available information is subject to varying levels of quality, detail, variability, and uncertainty. Nevertheless, the tool is far from unusable and can provide meaningful information for aiding decision-making, especially if it takes the levels of variability and uncertainty into account and thus avoids misleading results.

Cost-benefit analysis is used to determine whether an investment represents an efficient use of resources. An investment project regarding prevention represents an allocation of means (money, time, etc.) in the present that will result in a particular stream of hypothetical benefits in the future. The role of cost-benefit analysis is to provide information to the decision-maker, in this case an employee or a manager who will appraise the safety investment project. The main purpose of the analysis is to obtain relevant information about the level and distribution of benefits and costs of the project. Through this information, an investment decision within the company can be guided and made in a more objective way. The analysis role is to support objective evaluation and not to adopt an advocacy position either in favor of or against the safety investment, because there are also many other aspects that should be taken into account when deciding about safety investments (eg, social acceptability and regulatory affairs).

A safety investment project makes a difference, because the future will be different depending on whether the company decides to invest or not, or to invest in an alternative investment option. Thus, in the cost-benefit analysis, two scenarios are envisaged: without and with safety investment. During a cost-benefit analysis, a monetary value is assigned to the difference between the two scenarios. The process is shown in Fig. 16.1.

Because a safety investment project involves costs in the present and both costs and benefits in the future, at the beginning the net benefit stream will be negative; it will become positive at a certain point in time.

2.2 Present Value, Discount Rate, and Annuity

If a company uses a cost-benefit analysis, the recommendation whether to accept or to reject an investment project is based on the following process:
1. identification of costs and benefits
2. calculation of the present values of all costs and benefits
3. comparison of the total present value of costs and total present value of benefits

To compare the total costs and the total benefits (defined by costs and benefits occurring at different points in time), one needs to take a discount rate into account in the calculation to obtain the present values. Thus, during a cost-benefit analysis, all cash flows need to be converted to values in the present. This conversion is carried out by discounting the cash flows by a discount rate. The discount rate may be defined in a multiperiod model, where people value future experiences to a lesser degree than present ones because they are sure about present events and not sure about future events, which are subject to the environment. Thus, the higher the discount rate they choose, the lower the present values of the future cash flows [4].

Figure 16.1 Cost-benefit analysis approach for safety investments.

An investment project is recommended when the total net present value (NPV) of all cash flows is positive, and an investment project is usually rejected when the NPV is negative. To calculate the NPV related to project management, all cash flows are determined, and future cash flows are recalculated to today's value of money by discounting them by the discount rate. The formula usually mentioned to calculate the NPV is

$$NPV = \sum_{t=0}^{T} \frac{X_t}{(1+r)^t}$$

where X_t represents the cash flow in year t, T is the time period considered (usually expressed in years), and r is the discount rate.

Applied to operational safety, the NPV of a project expresses the difference between the total discounted present value of the benefits and the total discounted present value of the costs. A positive NPV for a given safety investment indicates that the project benefits are larger than its costs.

$$NPV = \text{Present value (benefits)} - \text{Present value (costs)}.$$

If $NPV \geq 0$, recommend safety investment.

If $NPV < 0$, recommend to reject safety investment.

It is evident that cash flows, that is, prevention costs and certainly hypothetical benefits, may be uncertain. Different approaches can be used in this regard. Cash flows can, for example, be expressed as expected values, taking the uncertainties in the form of probabilities into consideration and also increasing the discount rate to outweigh the possibilities for unfavorable outcomes.

There can be different categories of costs related to a safety investment, such as initial costs, installation costs, operating costs, maintenance costs, and inspection costs. These costs are evidently represented by negative cash flows. Some costs (eg, initial and installation costs) occur in the present and thus do not have to be discounted, whereas other costs (eg, operating, maintenance, and inspection costs) occur throughout the remaining lifetime of the facility and thus will have to be discounted to the present. There are also different categories of benefits (ie, avoided accident costs) linked to a safety investment, such as supply chain benefits, damage benefits, legal benefits, insurance benefits, human and environmental benefits, intervention benefits, reputation benefits, and other benefits. The benefits represent positive cash flows, which all occur throughout the remaining lifetime of the facility and thus will all have to be discounted to the present.

To clarify the discount rate principle, all cash flows (for both costs and benefits) are assumed to occur on an arbitrarily chosen date, which for example can be chosen to be the last day of the calendar year in which they occur. This assumption converts the continuous cash flows to a discrete range of cash flows, occurring at the end of each year. Then the cash flows at the end of each year have to be discounted to a present value, using a discount factor.

Suppose a safety investment project has a particular cost in year zero and then the same level of costs and benefits at the end of each and every subsequent year for the remaining lifetime of the facility. This means that:

$$\text{Cost}_i = C; \quad \text{Benefit}_i = B \quad \forall i \in T$$

This concept is called an annuity [4]. The total present value (TPV) of annuities is given by the following formula, with n the remaining lifetime of the facility

$$\text{TPV} = \frac{A[(1+r)^n - 1]}{[r(1+r)^n]} = A\frac{1 - (1+r)^{-n}}{r}$$

with A the yearly cost or benefit of a cost/benefit category, n the number of years in which the safety investment will be active, and r the discount rate. The term $\frac{[(1+r)^n - 1]}{[r(1+r)^n]}$ $\left(\text{or } \frac{1 - (1+r)^{-n}}{r}\right)$ is called the annuity (discount) factor [4].

Using this model, the benefits and costs in the future are assumed to be constant, and inflation is not included in the future costs and benefits, as already mentioned. Inflation is the process that results in a rise of the nominal prices of goods and services over time. Therefore in this (simplified) model, the real rate of interest[1] should be used as the discount rate instead of the money rate of interest. Because the money rate of interest m includes two components, the real rate of interest r and the anticipated rate of inflation i ($m = r + i$), the anticipated rate of inflation is built into the money rate of interest. Inflation is not included in the numerator of the formula for calculating the present value of annuities (because the costs and benefits are constant throughout the remaining lifetime), but it can be added.

2.3 Investment Analysis

Investment analysis allows calculating the internal rate of return (IRR) and the payback period (PBP), which are important determinants of whether to undertake a safety project and invest in safety measures.

2.3.1 Internal Rate of Return

IRR can be defined as the discount rate at which the present value of all future cash flows (or monetized expected hypothetical benefits) is equal to the initial investment, that is, the rate at which an investment breaks even. It can be used to measure and compare

---

[1] Real rate of interest (r): does not include the anticipated rate of inflation (i).

the profitability of investments. Generally speaking, the higher an investment's IRR, the more desirable it is to carry on with the investment. As such, the IRR can be used to rank several possible investment options that an organization is considering. Assuming all other factors are equal among the various investments, the safety investment with the highest IRR would be recommended to have priority. IRR is sometimes referred to as economic rate of return.

An organization should, in theory, undertake all safety investments available with IRRs that exceed a minimum acceptable rate of return predetermined by the company. Investments may of course be limited by availability of funds to the company. Because the IRR return is a rate quantity, it is an indicator of the efficiency, quality, or yield of an investment. This is in contrast with the NPV, which is an indicator of the value or magnitude of an investment.

The IRR $r*$ is a rate of return for which the NPV is zero. This can be expressed as follows:

$$\text{NPV}(r*) = \sum_{n=0}^{N} \frac{C_n}{(1 + r*)^n} = 0$$

In cases when a first safety investment displays a lower IRR but a higher NPV over a second safety investment, the first investment should be accepted over the second one. Furthermore, the IRR should not be used to compare investments of different duration. For example, a NPV of an investment with longer duration but lower IRR could be higher than a NPV of a similar investment (in terms of total net cash flows) with shorter duration and higher IRR.

2.3.2 Payback Period

PBP is defined by calculating the time needed (usually expressed in years) to recover an investment. Thus, a break-even point of investment is determined. The PBP of a certain safety investment is a possible determinant of whether to proceed with the safety project, because longer PBPs are typically not desirable for some companies. It should be noted that PBP ignores any benefits that occur after the determined time period and does not measure profitability. Moreover, neither time value of money nor opportunity costs are taken into account in the concept. PBP may be calculated as the cost of safety investment divided by the annual benefit inflows.

It is worth noting that PBP calculation uses cash flows, not the net income. PBP simply computes how fast a company will recover its cash investment.

3. PREVENTION COSTS

The purpose of implementing operational safety measures is to reduce present and future operational risks. Reducing risk implies both prevention of accidents and mitigation of potential consequences.

3.1 Safety Measures

There are four different classifications in which safety measures can be divided. First, risk reduction measures can be classified into protection and prevention measures, depending on their characteristics. Protection measures (including mitigation measures) lower the consequences, whereas prevention measures decrease the probability of an accident [5]. Second, safety measures can also be classified in active or passive systems according to their needs to be able to perform their function. Third, a classification can be made by their impact on severity and probability of occurrence, because from the many safety measures only some of them will, for example, play a role in the prevention of catastrophic or disastrous events. There may be a need to focus the efforts and to identify priority safety elements. Therefore, the third classification of safety measures is safety measures and safety-critical measures. Fourth, safety measures can be looked at from three dimensions: people, procedures, and technology. The interplay between people, procedures, and technology safety measures defines the observable part of the safety culture in an organization [6].

3.2 Categories of Safety Measure Costs

To be able to implement new safety measures and upgrade existing safety systems, a company has to reserve substantial funding. The various categories of possible costs related to new safety measures that a company has to take into account are are shown in Table 16.1. This table provides a clear overview of the different kinds of costs that may be applicable for safety.

4. HYPOTHETICAL BENEFITS: CATEGORIES OF AVOIDED ACCIDENT COSTS

Table 16.2 illustrates the avoided costs, also known as hypothetical benefits. These costs are grouped into eight categories depending on the type of benefits. Because the consequences of an accident only become reality when the accident actually occurs, the frequency of occurrence should be taken into account in the calculation of the expected events. Therefore, consequences, calculated by either the appropriate formula or a flat-rate amount, should be multiplied by the frequency of occurrence to obtain the "yearly" expected consequences resulting from an accident. Thus, if we consider that the different kinds of hypothetical consequences are to be spread out on a yearly basis and that the yearly cost resulting from these consequences is always the same, $C_i = C$, $\forall\ i$, the total present value of all hypothetical consequence costs during the remaining lifetime of the facility can be calculated by taking into account both the remaining lifetime and a discount factor.

   This calculation should be executed for both cases with and cases without the implementation of the safety measures. The difference between the two present values of

Table 16.1 Cost Categories of Safety Measures

Categories of Costs	Subcategories of Costs
Initial costs	Investigation Selection and design Material Training Changing guidelines and informing team
Installation costs	Production loss Start-up Equipment Installation
Operating costs	Utilities
Maintenance costs	Material Maintenance team Production loss Start-up
Inspection costs	Inspection team
Logistics and transport safety costs	Transport of hazmat Storage of hazmat Drafting of control lists Safety documents
Contractor safety costs	Team selection Training
Other safety costs	Other safety costs

consequence costs represents the (hypothetical) benefit resulting from the implementation of the new safety measures.

Benefits = Total present value of expected consequence costs (without safety investment) − Total present value of expected consequence costs (with safety investment)

5. COST-BENEFIT ANALYSIS

The following equation may be used to determine whether the costs of a safety measure outweigh its benefits [3]:

$$[(C_{without} \cdot F_{without}) - (C_{with} \cdot F_{with})] \cdot Pr_{control} > \text{Safety measure cost}$$

If information on the frequencies of initiating events is not sufficient

$$(C_{without} - C_{with}) \cdot F_{accident} \cdot Pr_{control} > \text{Safety measure cost}$$

with $C_{without}$ = cost of accident without safety measure, C_{with} = cost of accident with safety measure, $F_{without}$ = statistical frequency of initiating event if the safety measure

Table 16.2 Categories of Avoided Costs

Type of Benefits	Subcategory
Supply chain	Production loss 　Start-up 　Schedule
Damage	Damage to own material/property 　Damage to other companies' 　material/property 　Damage to surrounding living 　areas 　Damage to public material 　property
Legal	Fines 　Interim lawyers 　Specialized lawyers 　Internal research team 　Experts at hearings 　Legislation 　Permits and licenses
Insurance	Insurance premiums
Human and environmental	Compensation victims 　Injured employees 　Recruiting 　Environmental damage
Intervention	Intervention
Reputation	Share price
Other	Manager working time 　Cleaning

is not implemented, F_{with} = statistical frequency of initiating event if the safety measure is implemented, $F_{accident}$ = statistical frequency of the accident, and $Pr_{control}$ = probability that the safety measure will perform as required.

Accidents may be related to extremely low frequencies and a high level of uncertainty. In this case, these formulas cannot be applied because rough estimations would seriously compromise their reliability. To take this into account, cost-benefit analysis preferably involves a so-called disproportion factor to reflect an intended bias in favor of safety over costs. This safety mechanism is extremely important in the calculation to determine the adequate level of investment in prevention measures; on the one hand, the probability influences the hypothetical benefits substantially through the number of years over which the total accident costs can be spread out, and on the other hand, the uncertainty regarding the consequences is high [7].

Usually cost-benefit analyses state that the investment is not encouraged if the costs are higher than the benefits. However, if a disproportion factor is included, an investment in safety is reasonably practicable unless its costs are grossly disproportionate to the benefits. If the following equation is true, then the safety measure under consideration is not reasonably practicable because the costs of the safety measure are disproportionate to its benefits [8].

$$\text{Costs/benefits} > \text{Disproportion factor} \rightarrow E > \text{Benefits} \times \text{Disproportion factor}$$

To give an idea about the size of the disproportion factor, some guidelines and rules of thumb are available. They state that disproportion factors are rarely greater than 10, and that the higher the risk, the higher the disproportion factor should be to stress the magnitude of those risks in the cost-benefit analysis. This means that in cases when the risk is very high, it might be acceptable to use a disproportion factor greater than 10 [7]. Although a value greater than 10 is allowed, Rushton strongly advises not to use a disproportion factor greater than 30 [9].

6. CONCLUSIONS

In this chapter some concepts and methods borrowed from economic theory are used to assess and analyze investments in security. In common organizational practices, owing to the difficulties in quantifying the savings or benefits resulting from investments in security, the managerial focus is mainly on short-term objectives, postponing or delaying safety investment decisions. Moreover, the lack of appropriate methodologies to evaluate the economic impact of operational safety might be misleading, resulting in erroneous management solutions for an organization in the mid to long term.

Based on the information available in a company after an accurate risk assessment, cost-benefit analysis can be used to evaluate a safety measure and/or compare different safety alternative investments. A disproportion factor may be used in the case of high uncertainties in dealing with high-consequence, low-probability scenarios.

A cost-benefit analysis applied to safety measures can be used by decision-makers and/or analysts to provide recommendations for investments in prevention and make the evaluation phase more transparent and straightforward based on quantitative and homogeneous measures.

The concept of NPV and IRR can be used based on the expected yearly cash flow that an investment in safety generates. These concepts can potentially avoid myopic operational safety focused on the short term. Moreover, a disproportion factor is used to increase the risk awareness of the decision-maker and make the benefits comparable with the costs.

To conclude, operational safety can be reprioritized by using cost-benefit analysis described in this chapter, rebalancing the weight that managers give to production over safety and allowing decision-makers to take better management solutions by considering a broader investment horizon. Moreover, safety investments can be easily justified and evaluated using quantitative approaches.

REFERENCES

[1] Perrow C. Normal accidents: living with high risk technologies. Princeton University Press; 1984.
[2] Jackson D. Healthcare economics made easy. Scion Publishing; 2012.
[2a] Paltrinieri N, Bonvicini S, Spadoni G, Cozzani V. Cost-benefit analysis of passive fire protections in road LPG transportation. Risk Anal 2012;32:200–19.
[3] Boardman AE. Cost-benefit analysis: concepts and practice. Pearson/Prentice Hall; 2006.
[4] Campbell HF. Brown RPC. Benefit-cost analysis: financial and economic appraisal using spreadsheets. Cambridge University Press; 2003.
[5] Meyer T, Reniers G. Engineering risk management. De Gruyter; 2013.
[6] Reniers GLL. Multi-plant safety and security management in the Chemical and process industries. Wiley; 2010.
[7] Goose MH. Gross disproportion, step by step—a possible approach to evaluating additional measures at COMAH Sites. Bootle, UK: Health and Safety Executive; 2006.
[8] HSE Health and Safety Executive. Cost benefit analysis (CBA) checklist. 2013. http://www.hse.gov.uk/risk/theory/alarpcheck.htm.
[9] Rushton A. CBA, ALARP and industrial safety in the United Kingdom. UK. 2006.

CHAPTER 17

Cost-Benefit Analysis for Low-Probability, High-Impact Risks: Tutorials and Examples

L. Talarico[1], G. Reniers[1,2,3], N. Paltrinieri[4,5]

[1]University of Antwerp, Antwerp, Belgium; [2]KU Leuven, Brussels, Belgium; [3]Delft University of Technology, Delft, The Netherlands; [4]Norwegian University of Science and Technology (NTNU), Trondheim, Norway; [5]SINTEF Technology and Society, Trondheim, Norway

1. INTRODUCTION

The evolution of risk analysis into a dynamic approach and the use of sophisticated support techniques represent the response to the increasing need for modeling and assessing a wider range of potential accident events. Particular attention is being dedicated to low-probability, high-impact events, which are challenging not only to prevent but also to describe because of their rarity. Within this context, cost-benefit analysis (CBA) of safety investments, aimed at controlling this type of major accident, represents a challenging and crucial process. Because of the relative lack of data on these potential scenarios, not all the economic measures involved in the decision can be calculated with an acceptable level of accuracy, and the notions of cost and benefit need to be broadened.

On the one hand, costs should include not only the financial sacrifice demanded of an organization to purchase a safety measure but also the related criticalities and negative side effects (both quantitatively and qualitatively measurable). On the other hand, benefits do not necessarily imply monetary return of the investment but include the concepts of advantages, reputation, and respect of legal frameworks and policies. Although these elements are difficult to assess, recommendations to decision-makers can be provided by using different techniques of CBA.

A disproportion factor may be used as an adjustment in favor of safety when partial quantitative data about costs and benefits are available. When quantitative data are not available or when qualitative elements need to be considered in the decision-making process, a multicriteria decision-making approach may be used to rank and assess safety measures.

This chapter presents tutorials and application examples of both these approaches for CBA of safety measures.

Dynamic Risk Analysis in the Chemical and Petroleum Industry
ISBN 978-0-12-803765-2

2. COST-BENEFIT ANALYSIS BASED ON THE DISPROPORTION FACTOR

Thomas and Jones [1] presented an approach that is widely used within companies to assess alternative safety measures based on the maximum justifiable spend (MJS). MJS represents the amount of money worth spending to reduce risk to "acceptable" levels based on the magnitude of the risk itself. It provides a high-level criterion useful in identifying which additional safety measures are worth considering. According to this threshold, safety investments higher (more costly) than the MJS are to be discarded. Conversely, those interventions that cost less than the MJS can be recommended to be implemented. This method represents an alternative to the calculation of the net present value (NPV), which can be used within a CBA to shortlist alternative safety investments.

2.1 Methodology Tutorial

A methodological approach is shown in Fig. 17.1. It is composed by 4 steps in which a different type of assessment is carried out as explained below.

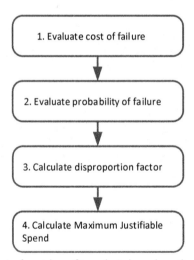

Figure 17.1 Flowchart of cost-benefit analysis based on the disproportion factor.

2.1.1 Step 1: Evaluate Cost of Failure

A worst-case accident scenario composed of a series of unwanted events should be identified. The analyst should assess costs of potential fatalities and disruptions to equipment and infrastructure caused by the events identified. Standard methodologies for this step are reported elsewhere [2,3]. See also Chapter 16 on CBA.

2.1.2 Step 2: Evaluate Probability of Failure

Failure probabilities for the events identified in the scenario should be assessed. Standard methodologies for this step are reported elsewhere [2,3].

2.1.3 Step 3: Calculate Disproportion Factor

Goose [4] suggests calculating the disproportion factor from the multiplication of the "how factors" and the addition of a dimensionless number to the product, as in Eq. (17.1):

$$\text{Disproportion factor} = \text{"how bad"} \times \text{"how risky"} \times \text{"how variable"} + 3 \qquad (17.1)$$

The "how bad" factor is computed considering the effects on the disproportion factor of the average number of fatalities (N_{av}) per event identified in the scenario, as in Eq. (17.2).

$$\text{"How bad"} = \log_{10}(N_{av}) \qquad (17.2)$$

The "how risky" factor represents the effect on the disproportion factor of the expectation value (EV, ie, the average number of fatalities expected per year, sometimes referred to as potential life loss), as in Eq. (17.3):

$$\text{"How risky"} = \log_{10}\left(10^5 \times EV\right) \qquad (17.3)$$

The "how variable" factor represents the effect on the disproportion factor of the ratio between the maximum potential number of fatalities (N_{max}) and the average number of fatalities (N_{av}) per event, as in Eq. (17.4):

$$\text{"How variable"} = \log_{10}\left(\frac{N_{max}}{N_{av}}\right) \qquad (17.4)$$

2.1.4 Step 4: Calculate Maximum Justifiable Spend

MJS is finally calculated as follows:

$$\text{MJS} = \text{cost of failure} \times \text{probability of failure} \times \text{disproportion factor} \qquad (17.5)$$

2.2 Application

This approach has been used to find reasonable safety budgets to be invested on the safety measures summarized in Table 17.1 for the prevention of domino effects in a chemical plant [5].

Because of the presence of dangerous and flammable chemical substances and the configuration of the plant, a team of risk experts has concluded that the plant may be exposed to vapor cloud explosions and large fires, which may trigger domino effects. In the worst-case scenario, such events may potentially lead to fatalities and disruptions

Table 17.1 Characteristics of Available Safety Measures

Measures	Purchase Cost (m €)	Effectiveness (%)
Concrete wall surrounding tank of 25 m + sprinkler without additional foam	15	95
Automatic sprinkler installation with additional foam	10	93
Automatic sprinkler installation without additional foam	8	92
Deluge system (water spray system opened as signaled by a fire alarm system)	4	90
Fire-resistant coating	2	84

to equipment and infrastructure quantified at €10 billion. The failure probabilities for such events range between 10^{-5} and 10^{-4}.

The calculation of the disproportion factor by means of Eq. (17.1) (not reported here for brevity) results in the suggestion to accept a safety investment five times higher than the potential benefits (ie, a disproportion factor of 5).

Using the MJS measure in Eq. (17.5), the maximum budget to be spent on safety may lie between €0.5 m and €5 m, depending on the likelihood of the failure. Only two of the five available options in Table 17.1 represent viable investment options. The deluge system (water spray system opened as signaled by a fire alarm system), with a cost equal to €4 m, should be selected owing to its higher effectiveness in preventing or mitigating domino effects.

3. COST-BENEFIT ANALYSIS BASED ON MULTICRITERIA DECISION-MAKING

As a result of a lack of data or high levels of uncertainty, in many real-life situations it is extremely complex to carry out a full CBA. As a result, a high-level analysis can be conducted based on a general CBA including both quantitative and qualitative evaluation elements. More specifically, a multicriteria analysis (MCA) may allow decision-makers to compare different alternative safety investments. MCA can be used for many purposes: ranking options, identifying the most suitable investment, or detecting a limited number of options for more detailed appraisal. MCA represents a powerful methodology to support decision-making and improve the quality of the resulting decisions.

Differently from pure CBA—which includes methods based on the NPV considering only economic aspects quantifiable in monetary terms—MCA generalizes a traditional CBA, also including in the assessment other, not purely economic measures, if these are considered useful in the decision process by decision-makers.

To give more emphasis to economic factors such as costs and benefits, one might allocate a pivotal role to these criteria (eg, by associating high weights to them, as explained later), to significantly influence the decision-making and be in line with traditional CBA methodologies. More details are reported elsewhere [6–8].

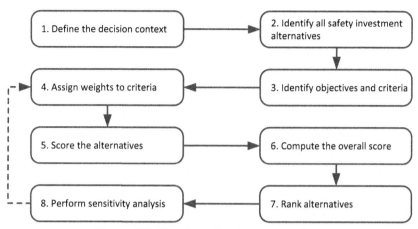

Figure 17.2 Flowchart of multicriteria analysis.

3.1 Methodology Tutorial

A structured MCA methodology is characterized by the steps shown in Fig. 17.2.

3.1.1 Step 1: Define the Decision Context

The first step involves the definition of the decision context: system considered, final goals of the analysis, and decision-makers involved in the decision process.

3.1.2 Step 2: Identify All Safety Investment Alternatives

Alternative investments should be identified and compared to a "no-investment" base case. Differences are to be consistently measured over a clearly defined timeline depending on the duration′ of the assets or technologies involved.

3.1.3 Step 3: Identify Objectives and Criteria

Objectives and relevant criteria for the evaluation should be identified. Criteria should reflect the values associated with the impact of every alternative/option. They may reflect critical issues that need to be assessed by decision-makers.

3.1.4 Step 4: Assign Weights to Criteria

A weight can be associated with each criterion to reflect the relative importance of one criterion over the others, according to the decision-makers.

3.1.5 Step 5: Score the Alternatives

Once this framework has been set, the evaluation of the alternatives can be performed by assigning values (scores) to each alternative for each of these criteria based on a predefined scale.

3.1.6 Step 6: Compute the Overall Score

Weights and scores are combined through a weighted sum to assign a total weighted score for each investment.

3.1.7 Step 7: Rank Alternatives

The alternative investments are compared and ranked to identify which are the options that deserve further investigation.

3.1.8 Step 8: Perform Sensitivity Analysis

Finally, a sensitivity analysis may be performed to assess whether a change in score or in weight, associated with specific criteria, significantly affects the final rank.

3.2 Application

The MCA approach used most commonly is the linear additive model, which has benefits in its simplicity and user friendliness. In this example, MCA is applied by a public railway transport organization, which has to evaluate complex safety investments aimed at preventing and mitigating major accidents. MCA has been adopted owing to scarcity of historical data, lack of accurate measures for costs and benefits expressed, and need to consider qualitative indicators in the analysis.

The goal is to evaluate alternative investments for train control systems aimed at increasing the overall level of safety in the entire railway network. The level of safety is measured considering the number of trains that are traveling on the railway network, the average number of passengers on each train, the nature of the materials transported (for freight transport), the environment, and the population living next to the railway lanes, which may be affected in case of a train accident.

Given the current railway network and its related level of safety, the firm has received by the government a mandate to improve by 20% the level of safety in a time horizon of 5 years without exceeding a predefined safety budget. Some improved safety technologies can be installed both on trains and on network infrastructures to achieve this goal. However, each safety technology presents different features such as installation costs, maintenance costs, effectiveness (expressed as the capability to improve the current safety level), compatibilities with other technologies and systems, and duration of the installation. The no investment case is represented by the current state of the overall railway network. Owing to the complexity of the project, a team of experts from different departments has defined eight alternative investments, allowing increase of the overall safety level by using different approaches. In fact, these options present different levels of expenditure and require specific changes to the system. Moreover, each option has a different timing structure, specific implementation difficulties, and criticalities.

The main features of an improved integrated railway network (made by network infrastructure and rolling stock) are essentially related to investment costs, yearly recurrent costs, safety level, compatibility with other ongoing modernization projects, railway capacity requirements, project duration, and implementation risks. Because these features represent critical issues that need to be assessed by decision-makers, they can be treated as criteria to be used to evaluate each investment option. Fourteen criteria have been identified, which correspond to the set of relevant objectives to be pursued by decision-makers. Their evaluation will facilitate the comparison of the alternative investment options. The list of selected criteria, their description, the evaluation metrics, and the related weights are summarized in Table 17.2.

Weights are associated with the criteria, reflecting their relative importance and making sure that the sum of the weights is equal to 100%. Criterion number 1 has been excluded because it represents a showstopper that may spoil the whole investment if it is not compliant with transport legislations.

Because all these criteria cannot be properly quantified, marginal expected differences between alternative projects and the no investment scenario are assessed using a standard scale defined through expert judgment. This score is useful in ranking each alternative. Performances are scored using a Likert scale (except for criterion 1) from 1 to 5, where 1 represents a low (generalized) benefit or high (generalized) cost, and 5 indicates a high (generalized) benefit or low (generalized) cost.

Given the list of weights w_i and metrics (with a minimum and maximum score s_i) for each criterion i, the total weighted score $\left(\sum_{i=1}^{13} w_i \cdot s_i\right)$ associated with a safety investment option should lie in the interval between 100 and 500. For instance, if an option scores 1 on all criteria, the total weighted score is 100. At the other extreme, the maximum total weighted score for an investment assumes the value of 500 if the project scores 5 on all criteria.

In Table 17.3 the scores associated with each safety investment are shown. Investment option H does not respect the requirement to be compliant with legislations on railway transport and for this reason is disregarded. The remaining seven investments (from A to G) are ranked.

Alternative investment C appears to be the best choice. It scores 441 of a maximum of 500. In light of this result, the recommendation that the decision-makers can provide is that investment option C deserves further investigation. No sensitivity analysis was performed to assess the validity of the first-ranked solution.

Table 17.2 Description of the Criteria Used in Multicriteria Analysis

ID	Criteria	Description	Metric	Weight (%)
1	Transport regulations	The investment option is compliant with all the transport regulations. This criterion represents a clear showstopper.	Yes/no	—
2	Sustainability	The investment is feasible and sustainable from a technical, financial, and economical point of view. Moreover, its impact on the environment is limited and within the boundaries defined by law. The higher the score, the more sustainable the option.	Score (1–5)	10
3	Duration	The investment can be completed by the due date (project duration 5 years). A high score for this criterion means that the project can be completed even before the deadline.	Score (1–5)	5
4	Initial investment costs	The expected capital investment required by the alternative investment. Assuming that the total investment costs do not exceed the initial safety budget, the higher the score, the lower the economic burden associated with the investment.	Score (1–5)	15
5	Yearly recurring costs	The expected recurring costs required by each investment after its completion and incurred on a yearly basis due eg, to maintenance activities, energy consumption, insurance premium. The higher the score, the lower the capital required by the investment on a yearly basis.	Score (1–5)	10
6	Safety level	The improvement level that can be achieved by the investment owing to the installation of enhanced safety technologies on both rolling stock and network infrastructure. Assuming that all alternatives guarantee the minimal safety increase threshold of 20%, the higher the score, the higher the expected safety improvement that can be achieved.	Score (1–5)	20
7	Capacity improvement	The total number of trains that can travel on the railway network after the implementation of the investment compared with current values. The higher the score, the larger the improvement guaranteed by the investment.	Score (1–5)	5

8	Compatibility with existing projects	The investment does not present any incompatibility with existing investment projects. The higher the score, the lower the level of incompatibility with ongoing projects.	Score (1–5)	4
9	Availability and reliability	The robustness of the comprehensive railway network after the modernization investment. The higher the score, the better the investment.	Score (1–5)	4
10	Risk to delivery	The likelihood that the project will miss the expected due date. The higher the score, the lower the risk of project delays.	Score (1–5)	7
11	Implementation difficulties	Level of criticalities and foreseeable problems that might show up during the implementation phase. The higher the score, the higher the level of confidence for a smooth implementation.	Score (1–5)	2
12	Lanes affected	Number of network lanes affected by the investment projects owing to modernization and/or installation of new technologies. The higher the score, the lower the amount of network infrastructure affected by the investment.	Score (1–5)	6
13	Rolling stock affected	Number of trains affected by the investment project owing to modernization and/or installation of new technologies. The higher the score, the lower the amount of rolling stock affected.	Score (1–5)	7
14	External risk	Risk related to the trustworthiness and/or other criticalities that might arise with suppliers, partners, and/or third parties during the implementation of the investment. The higher the score, the lower the expected external risk the investment will face during implementation.	Score (1–5)	5

Table 17.3 Evaluation of Alternative Investments

Criteria	Weight (%)	Alternative Investments							
		A	B	C	D	E	F	G	H
Transport regulations	—	Yes	Yes	Yes	Yes	Yes	Yes	Yes	No
Sustainability	10	1	3	5	4	2	3	2	
Duration	5	2	5	4	3	2	4	2	
Initial investment costs	15	5	5	4	3	2	3	5	
Yearly recurring costs	10	4	4	5	5	3	5	4	
Safety level	20	4	4	5	4	4	3	4	
Capacity improvement	5	2	3	3	4	5	3	2	
Compatibility with existing projects	4	2	3	5	4	5	4	3	
Availability and reliability	4	5	5	4	4	4	2	4	
Risk to delivery	7	1	4	5	5	5	5	5	
Implementation difficulties	2	5	3	4	4	4	3	4	
Lanes affected	6	3	5	4	4	3	1	5	
Rolling stock affected	7	3	5	4	4	4	5	4	
External risk	5	5	4	3	3	4	4	5	
Total weighted score	**100**	**334**	**416**	**441**	**392**	**340**	**346**	**389**	
Rank		**5**	**2**	**1**	**3**	**7**	**6**	**4**	

4. CONCLUSIONS

CBA can be effectively used within a more robust risk management phase to support decision-makers. For complex investments, such as those involving sophisticated technical equipment, infrastructures, high level of uncertainty, and variability of data, different approaches may be applied.

In some cases, costs and benefits associated with investments are difficult to estimate, the probability associated with an accident scenario may be low, and NPV quantitative assessment might not be applicable. For this reason, a disproportion factor in favor of safety is used.

Moreover, criteria associated with costs and benefits may be needed to rank and classify alternative safety investments. A quantitative CBA is still possible if costs and benefits are envisioned from a wider perspective, in which disadvantages and advantages associated with safety measures are considered. In this case, a multicriteria approach may be used to support decision-makers and improve the effectiveness of the safety management process.

REFERENCES

[1] Thomas PJ, Jones RD. Extending the J-value framework for safety analysis to include the environmental costs of a large accident. Process Safety and Environmental Protection 2010;88:297–317.

[2] Center for Chemical Process Safety. Guidelines for chemical process quantitative risk analysis. New York (USA): American Institute of Chemical Engineers (AIChE); 2000.

[3] TNO. Guidelines for quantitative risk assessment: purple book. den Hague (The Netherlands): National Institute of Public Health and Environment (RIVM); 2005.

[4] Goose MH. Gross disproportion, step by step-a possible approach to evaluating additional measures at COMAH sites. Institution of chemical engineers symposium series. Rugby (UK): Institution of Chemical Engineers; 2006.

[5] Janssens J, Talarico L, Reniers G, Sörensen K. A decision model to allocate protective safety barriers and mitigate domino effects. Reliability Engineering and System Safety 2015;143:44–52.

[6] Brans J-P, Mareschal B. Promethee methods. Multiple criteria decision analysis: state of the art surveys. New York: Springer; 2005. p. 163–86.

[7] Talarico L, Sörensen K, Springael J. A biobjective decision model to increase security and reduce travel costs in the cash–in–transit sector. International Transactions in Operational Research; 2015. http://dx.doi.org/10.1111/itor.12214.

[8] Talarico L, Reniers G, Sörensen K, Springael J. Intelligent systems in managerial decision making. In: Kahraman C, Çevik Onar S, editors. Intelligent techniques in engineering management. Springer International Publishing; 2015. p. 377–403.

SECTION 3.3

Reputation

CHAPTER 18

Reputational Damage After Major Accidents

K. Kyaw[1], N. Paltrinieri[2, 3]

[1]Trondheim Business School, Norwegian University of Science and Technology (NTNU), Trondheim, Norway;
[2]Norwegian University of Science and Technology (NTNU), Trondheim, Norway; [3]SINTEF Technology and Society, Trondheim, Norway

1. INTRODUCTION

When the Deepwater Horizon oil rig operated by British Petroleum exploded on April 20, 2010, the company incurred costs beyond any prediction. Most measures of costs correspond to those arising from damages to the facilities, cleaning up the environment or meeting liability claims from various parties. Those costs can be estimated with reasonable accuracy, and most attention focuses on them. Costs associated with damaged reputation, on the other hand, are difficult to estimate and thus are often neglected. Yet it is a fact that the company lost market value because of lost sales to customers and investors' diminished confidence in the company. Chapter 19 provides a detailed example on how the cost of reputational damage is estimated for the explosion at Deepwater Horizon oil rig. This chapter introduces ways to assess the cost of reputational damage due to past events and the related advantages and disadvantages. Such an estimation may provide detailed information that can be integrated into cost—benefit analyses of safety measures (Chapters 16 and 17) and represent an important caveat for risk analysts.

2. THEORETICAL BACKGROUND

2.1 Efficient Market Hypothesis

A market is information efficient if prices in the market adjust rapidly to the arrival of new information. A market can be information efficient in three forms: weak form efficiency, semistrong form efficiency, and strong form efficiency. Fama [1,2] provides an overview of the various forms of market efficiency and the corresponding empirical tests and results.

1. The weak form of the efficient market hypothesis (EMH) posits that current prices fully reflect all historical market information, such as historical trading volume, prices, and rates of returns.

2. The semistrong form of the EMH asserts that current prices reflect all publicly available information. Thus, by definition, semistrong form efficiency encompasses weak form efficiency because it reflects all market and nonmarket information that is publicly available. Public information includes any information that has relevance to a company's

earnings that is available to the general public. This may be earnings or dividend announcements made by a company; inflation and economic forecasts provided by the government; price-to-earnings ratios; or dividend-yield information provided by the analysts and so forth.

3. The strong form of the EMH hypothesizes that current market prices fully reflect all information that is available both publicly and privately. Because the current market price encompasses all information, it is a superior form of market efficiency than the semistrong form.

The question of whether a market is information efficient has been of vital interest to market participants. Academics have studied this for decades. For example, to test if a market is weak form efficient, studies have examined the predictability of futures prices based on past market information or investigated the statistical independence of historical prices. To test for semistrong form information efficiency, studies often examine the reaction of market prices to information that has been newly made public, such as the announcements of macroeconomic news or company events. To test for strong form information efficiency, studies focus on the ability of investment managers to generate above-average profits compared with ordinary investors. For instance, Bokhari et al. [3] test weak form market efficiency, Bernard and Thomas [4] examine semistrong form market efficiency, and Malkiel [5], Gruber [6], and Kosowski et al. [7] study strong form efficiency. Moreover, Fama [1] gives a good overview of the studies undertaken to test market efficiency. Those studies, among many others, generally conclude that developed markets are mostly weak form efficient but not strong form efficient, whereas studies based on announcements show a strong evidence of semistrong form efficiency.

3. ESTIMATING THE COST OF REPUTATIONAL DAMAGE

Several methods are used to predict the economic consequences of major accidents; Chapters 16 and 17 report some representative examples. However, predicting the cost of reputational damage is a challenging task that needs a solid and relevant basis.

For instance, Reniers and Brijs [8], who have developed a toll for intelligent allocation of prevention investments, consider the reputation consequences due to a major accident to be the expected loss in the market value of a company. The authors state that this estimation is based on the literature and expert opinion; thus, the study of past relevant accidents and the assessment of the resulting reputational damage would allow a more solid prediction of the potential unwanted consequences from an event.

A loss in reputational value can manifest itself in more than one way, and its assessment in the aftermath of a major accident is not as straightforward as it may appear. However, in a market where information is reasonably accurately reflected in the prices, it is reasonable to assume that events such as major accidents, which can negatively affect the earnings of a company, will be reflected in the company's share prices. It is on this assumption that the following methods were developed.

3.1 Net Worth Approach

In this approach, the value of a company to the shareholders as reported on the balance sheet of the company is compared with the market value of the shares. A company that has a good reputation will have shares with a much higher market value than the value reported on the balance sheet, suggesting that the net worth reported on the books is of good quality. The difference in the values will change as the company's reputation is damaged or built up. For example, when a company becomes more reputable, market value of the company equity will increase in relation to the net worth reported on the balance sheet. Thus, a company's reputation can be measured as

$$V_{CR} = MV - NW \tag{18.1}$$

where V_{CR} = value of company's reputation; MV = market value of a company = market price per share × number of shares; NW = net worth = total assets − total liabilities.

This approach is intuitive and easy to apply. However, the result may be inaccurate for several reasons. First, the balance sheet provides information about a company at a particular point in time, that is, on the date the statement is reported (t in Fig. 18.1). Therefore, the net worth value of a company reported on the balance sheet is representative of the value on that date or a few days around the date. As the company generates profits or losses from the business, the value of net worth will change ($t + 1$ in Fig. 18.1). Moreover, the price per share is subject to volatility. Thus, the value of reputation as

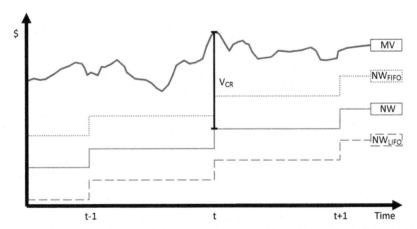

Figure 18.1 Graphic representation of the net worth approach and the related limits. *MV*, market value; *NW*, net worth; *NW_FIFO*, net worth under first in, first out principle; *NW_LIFO*, net worth under last in, first out principle; *t*, time; *V_CR*, value of company's reputation.

calculated in Eq. (18.1) is representative only of the value on date (t) when the balance sheet is reported.

Second, the value of reputation calculated from Eq. (18.1) is subject to accounting reporting principles because of the assets considered. For instance, the inventories may be reported using last in, first out (LIFO) principle or first in, first out (FIFO) principle. In LIFO, it is assumed that those inventories that are bought last are sold first; thus, the inventories listed on the balance sheet are reported at an earlier purchase price rather than the later purchase price. On the contrary, under FIFO, inventories are reported on the balance sheet at the most recent purchase price. A consequence of this is that the asset values may be overstated under the FIFO approach and, most likely, understated under the LIFO approach in a normal inflationary economic environment, altering the value of reputation obtained from Eq. (18.1). Moreover, when different companies adopt different reporting principles, it makes the comparison of the reputational values across companies difficult. Finally, accounting principles (eg, the Generally Accepted Accounting Principles (GAAP)) require an asset to be reported on the balance sheet at its purchase price. This is a disadvantage that is particularly relevant for companies whose balance sheet consists of large-value, long-term assets, such as an oil company. For example, an oil company usually has assets, such as the resource rights for a certain oil field, that may have been bought several years ago when prices were relatively lower. In such a case, the price of the assets at the time of purchase would be significantly lower than the price of the assets today, overestimating the company's reputation value. Fig. 18.1 summarizes the effects of different accounting principles on the value of net worth estimated from the balance sheet. Moreover, it shows the static nature of the net worth value in relation to the market value, which is dynamic over time.

3.2 Event Study Approach

Semistrong form market efficiency asserts that all relevant public information is incorporated into the current share price. As a result, share price will adjust as events in a company unfold. For instance, when a company announces an increase in quarterly earnings, share price will rise appropriately to reflect the positive news. Because the prices reflect the newly arrived public information, it is possible to extract the cost of reputational damage by using an event study approach. This approach had been widely applied in studying the reputational loss for a bank when it announced negative operating profits [9–11]. The approach measured the loss in the market value of the bank in excess of the realized operating losses as the cost associated with the loss in the reputation of the company. Later the approach was applied more widely to other types of major losses, such as major accidents experienced by industrial companies [12].

In a semistrong form of efficient market, returns on a share are expected to reflect publicly available information. When a company experiences a major accident, the market will revalue the company shares to reflect the effect of the accident on the current and future earnings of the company. Thus, a change in the market value of the company should correspond to the reduction in the earnings incurred as a result of the physical losses caused by the accident. Any reduction in the market value (often measured as negative returns on the shares) that is in excess of what is expected to result from an event can be attributable as an effect the event has on the company's reputation. The event study methodology offers a means to isolate the effect of an accident on the price of a stock. The resulting effect is then compared with the effect expected to have occurred from the physical losses. The difference is suggestive of the result of nonphysical losses, such as reputational damage. Fig. 18.2 depicts this approach.

Capital market theory studies the dynamics of financial markets. One of the most popular capital market theories is the capital asset pricing model. This model posits that in equilibrium, return on a security should commensurate with the level of market risk exposure to which an investor of the security is exposed. It argues that idiosyncratic risks such as firm-specific risk can be diversified away; thus, only nondiversifiable risk of a security should be rewarded with a return. The nondiversifiable risk, also known as the market risk, can be measured as a share variance attributable to overall market fluctuations [5,13—16]. Market risk of a share is measured by market beta, which is a share covariance to the overall market fluctuations, cov_{im}, standardized by market volatility, σ_m^2. Eq. (18.2) shows the formula to calculate market beta of a security.

1. Capital asset pricing model defines market beta as

$$\beta_i = \frac{cov_{im}}{\sigma_m^2} \tag{18.2}$$

where β_i = market risk as measured by market beta of security i; cov_{im} = covariance of a stock's returns to market returns; and σ_m^2 = variance of market returns. Covariance of stock returns to market returns, cov_{im}, is calculated as

$$\sigma_{im} = \sum_{t=1}^{n} \left(r_i - \overline{r_i} \right) \left(r_m - \overline{r_m} \right) \Big/ n$$

where r_i = return on a security i at time t; $\overline{r_i}$ = security i average return during period; $\overline{r_m}$ = market return at time t; r_m = market average return during period; and n = number of observations.

Variance of market returns, σ_m^2, is calculated as

$$\sigma_m^2 = \left(r_m - \overline{r_m} \right)^2 \Big/ n$$

where r_m = market return at time t and $\overline{r_m}$ = market average return during period.

2. Sharpe [17] proposed an empirical model to estimate the market beta using a market model. The market model is a statistical model that relates the return on a security to the market portfolio. In the capital asset pricing model, expected returns on a security

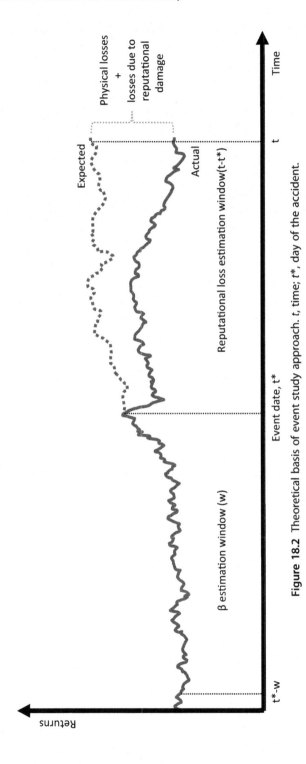

Figure 18.2 Theoretical basis of event study approach. t, time; t^*, day of the accident.

increase linearly with the security beta. Therefore the market model estimates a linear relationship between the market and the security returns and assumes a joint normality of asset returns. The model is defined as

$$r_{it} = \beta r_{mt} + \varepsilon_{it} \tag{18.3}$$

$$E[\varepsilon_{it}] = 0 \text{ and } \text{Var}[\varepsilon_{it}] = (\sigma_{\varepsilon i})^2$$

where r_{it} = returns on a company i at time t; r_{mt} = returns on the market at time t; ε_{it} is a disturbance term; ε_{it} is normally distributed with zero mean, $E[\varepsilon_{it}] = 0$; and Var $[\varepsilon_{it}] = (\sigma_{\varepsilon i})^2$ is constant variance,

In the application of Eqs. (18.2) and (18.3), a broad-based stock index is used as the market. For the United States, the Standard & Poor 500 index, the Center for Research in Security Price's value-weighted index, or the Dow Jones Industrial Average can be used, depending on how representative the index is as the market for the company being studied. For example, the Dow Jones Industrial Average may be an appropriate representative market for a blue chip company. Abnormal returns are then obtained from the forecast errors from the regression. Eq. (18.3) gives the beta estimate, $\widehat{\beta}$, which indicates how much the company share returns are expected to covary with the market returns. For instance, a beta estimate of 1.2 suggests that company returns will be 1.2 times the market returns. Thus, for a market return of 3%, the company returns are expected to be 3.6%, whereas they are expected to be -1.8% if market return is -1.5%. The beta estimate allows an estimation of (normal) returns that would be expected if an accident (ie, an event) did not take place. Thus, the abnormal returns to the company i on day t (AR_{it}) can be defined as

$$AR_{it} = r_{it} - \widehat{\beta} r_{mt} \tag{18.4}$$

where r_{it} = return for the company i on day t; r_{mt} = corresponding market return; and $\widehat{\beta}$ = estimated regression coefficient from Eq. (18.3).

The abnormal returns in Eq. (18.4), although abnormal in relation to general market conditions, are reflective of the specific major events affecting the company. Abnormal returns can also be estimated by other approaches, such as the Fama—French three-factor model. In the model, a stock return is adjusted for three factors: (1) market returns, (2) firm size, and (3) growth potential. This measure of abnormal returns is commonly applied in cross-sectional studies. Further details on the three-factor model are reported elsewhere [18].

The cost due to reputational loss during a period can then be defined as

$$RL_{it-T} = CAR_{it-T}(MV_{t*}) + AE_{it} \tag{18.5}$$

where RL_{it-T} = reputational loss for company i during the period $t - T$; CAR_{it-T} = cumulative abnormal returns for company i during the period $t - T$, which is calculated as $\sum AR_{it}$; MV_{t*} = market value of the company on the day of the accident t^*; and AE_{it} = actual expenses incurred by the company i on day t where $t \geq t^*$.

One major challenge in applying Eq. (18.5) to major accidents in industrial companies is that the AE_{it} figures are not immediately available. It may take months to years before an estimate of AE_{it} becomes available, and even longer for the corresponding actual AE_{it}. To accommodate for the possible time lapse between an accident and when the actual cost amount becomes known, Eq. (18.5) should be modified. Because the market would have begun to adjust once the news of an accident broke out, market value would have partially adjusted to the accident by the time the actual cost amount becomes available. Therefore, Way et al. [12] propose that the cost of reputational damage at time t (where t is n days after an accident) can be estimated as

$$RL_{it} = MV_{it} - MV_{it^*-1} - AE_{it} \tag{18.6}$$

where RL_{it} = reputational loss for company i on day t after the accident; MV_{it^*-1} = market value of the company on the day before the accident, where t^* is the day of the accident; MV_{it} = market capitalization of the company on day t, where $t \geq t^*$; and AE_{it} = expenses incurred by the company on day t, where $t \geq t^*$.

One major advantage of this approach over the net worth approach is that it does not rely on book value figures from the balance sheet in estimating reputational cost. One major disadvantage of the model is the reliance on the availability of actual cost figures. This can lead to even more difficulty if the company in question experiences other major accidents during the time lapse between the major accident and when the actual cost figures for the accident become known.

For example, at the end of 2005, an overfilling of unleaded petroleum in one of the storage tanks in the Buncefield (UK) storage depot led to the formation of a flammable vapor cloud, which dispersed inside the plant and to the surrounding facilities. As soon as the vapor cloud came into contact with an ignition source—which was believed to have been in the fire pump house, in the generator cabin, or from a car engine—a vapor cloud explosion of unexpected strength was generated. The explosion destroyed a large part of the depot, damaged the surrounding properties, and disrupted local communities. It is estimated that the accident cost 1.2 billion euros [19]. Although five companies were charged with offenses by the investigation committee for the Buncefield incident, Total UK Limited was considered the main party responsible for the accident. The accident occurred on December 11, 2005, yet the court decision on the amount that Total SA, the parent company of Total UK Limited, was liable to pay was made on September 24, 2012, almost seven years after the accident.

4. CONCLUSIONS

The net worth approach and the event study approach can be used to estimate the cost associated with reputational damage to a company in the event of a major accident. Such an assessment represents an important input in analyzing the economic consequences of a potential accident scenario, allowing for a general increase in the awareness of company risk and better support to safety-critical decision-making.

Although the net worth approach allows a quick approximation of the cost, it is relatively simplistic and inherits many pitfalls. The event study approach, on the other hand, provides more comprehensive and robust information about the cost. However, it is not flawless. In the event study approach, the cost of reputational damage is estimated as the loss in market value of a company after an accident in relation to the actual expenses incurred in the accident. This estimate is underpinned by market efficiency theories and has gained popularity in recent years in studying the reputational damage to industrial companies that have experienced major accidents, although it was initially used to study financial institutions that had experienced poor operating income.

REFERENCES

[1] Fama E. Efficient capital markets: a review of theory and empirical work. Journal of Finance 1970;25: 383—417.
[2] Fama E. Efficient capital markets: II. Journal of Finance 1991;46:1575—617.
[3] Bokhari J, Cai C, Hudson R, Keasey K. The predictive ability and profitability of technical trading rules: does company size matter? Economics Letters 2005;86:21—7.
[4] Bernard VL, Thomas JK. Post-earnings-announcement drift: delayed price response or risk premium? Journal of Accounting Research 1989;27:1—36.
[5] Malkiel BG. Returns from investing in equity mutual funds 1971 to 1991. Journal of Finance 1995;50: 549—72.
[6] Gruber MJ. Another puzzle: the growth in actively managed mutual funds. Journal of Finance 1996; 51:783—810.
[7] Kosowski R, Timmerman A, Wermers R, White H. Can mutual fund 'stars' really pick stocks? New evidence from a bootstrap analysis. Journal of Finance 2006;61:2551—95.
[8] Reniers G, Brijs T. Major accident management in the process industry: an expert tool called CESMA for intelligent allocation of prevention investments. Process Safety and Environmental Protection 2014;92:779—88.
[9] Fiordelisi F, Soana M, Schwizer P. The determinants of reputational risk in the banking sector. Journal of Banking & Finance 2013;37:1359—71.
[10] Gillet R, Hubner G, Plunus S. Operational risk and reputation in the financial industry. Journal of Banking & Finance 2010;34:224—35.
[11] Sturm P. Operational and reputational risk in the European banking industry: the market reaction to operational risk events. Journal of Economic Behavior & Organization 2013;85:191—206.
[12] Way B, Khan F, Weitch B. Is reputational risk quantifiable?. In: The international conference on marine safety and environment; 2013.
[13] Mossin J. Equilibrium in a capital asset market. Econometrica 1966;34:768—83.
[14] Sharpe W. Capital asset prices: a theory of market equilibrium under conditions of risk. Journal of Finance 1964;19:425—42.
[15] Treynor J. Market value, time, and risk. Unpublished manuscript. Rough Draft; 1961. p. 95—209.
[16] Treynor J. Toward a theory of market value of risky assets. In: Korajczyk R, editor. Asset pricing and portfolio performance. London: Risk Books; 1999. p. 15—22.
[17] Sharpe W. A simplified model for portfolio analysis. Management Science 1963;9:277—93.
[18] Fama E, French K. Size and book-to-market factors in earnings and returns. Journal of Finance 1995; 50:131—55.
[19] Buncefield Major Incident Investigation Board. The Buncefield incident 11 December 2005. The Final Report of the Major Incident Investigation Board. Bootle, UK: HSE Books; 2008.

CHAPTER 19

Estimation of Reputational Damage from Major Accidents: Tutorial and Examples

K. Kyaw[1], N. Paltrinieri[2,3]
[1]Trondheim Business School, Norwegian University of Science and Technology (NTNU), Trondheim, Norway;
[2]Norwegian University of Science and Technology (NTNU), Trondheim, Norway; [3]SINTEF Technology and Society,
Trondheim, Norway

1. INTRODUCTION

The cost of reputational damage can be measured using a net worth approach or an event study approach. A net worth approach compares the company's book value of equity with the market value of equity before and after an accident. Despite the ease with which the approach may be applied, it has its pitfalls, as mentioned in Chapter 18. A more accurate estimate is provided by the event study approach, which is described more in detail in the next section. In this chapter, both approaches are applied to the case of the major explosion that occurred at the Deepwater Horizon oil rig run by British Petroleum (BP) on April 20, 2010.

2. METHODOLOGY TUTORIAL

Fig. 19.1 gives a summary of the event study approach.

2.1 Prerequirements

There are two key pieces of information required to apply this approach. First, information about the costs incurred during an accident should be available. For a major accident, where damage is extensive, an actual cost amount may not be known until some months or years have elapsed. In such cases, an estimate may be used instead. Second, the company that has been affected by an accident should have its market value available or possible to be estimated. This prerequirement restricts the application of the approach to accidents experienced by publicly listed companies.

2.2 Step 1: Collect Information

To estimate the market beta and consequently the abnormal returns (ARs), historical price or return information on the company and the market should be collected first. Company market beta is estimated as the covariability of returns between the company

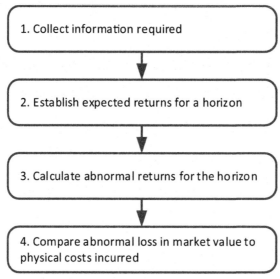

Figure 19.1 Event study approach.

and the market. Thus, historical price or return information should date back months before the accident. It is well documented that company beta can change over time. Thus, it is arguable that both excessively long and excessively short estimation periods can create a misrepresentation of the beta estimated. It is a common practice to use returns over the past 250 trading days to estimate the beta.

In addition to the market information, the actual costs incurred in an accident or the estimated costs should be available. The event study approach suggests that any abnormal loss in the market value above and beyond the actual costs incurred indicate the costs of reputational damage. Table 19.1 summarizes the most commonly used sources of information on the cost figures.

Table 19.1 Sources of Information Regarding the Actual and Estimated Costs of Physical Damage

Source	Remarks
News search	Internet searches yield the information that has made it to the news. The amount of information available can be limited but timely as the information about an accident unfolds.
Accident report	This is a report prepared by an official investigation bureau. The report usually contains information regarding the actual costs incurred from the date of the accident to the time that the report is prepared.
Annual report	This is a yearly report that a company has prepared for stakeholders, such as investors and lenders. The report summarizes the yearly and cumulative costs incurred by the accident.

2.3 Step 2: Establish Expected Returns for a Horizon

First, estimate the company beta from the estimation window. The estimation window should end just before the day of the particular accident, unless there have been other major events that may interfere with or distort the beta estimate. Then estimate the returns that would have been expected to prevail during a time horizon if there were no accident. The choice of the length of the horizon depends on the availability of the cost figures, among others. For instance, after the accident, an estimate of the costs may not be available until a few days have passed. In this case, the expected returns are calculated to the day when the cost figures, either actual or estimate, become available. The two steps to establish expected returns can be summarized as in Table 19.2.

2.4 Step 3: Calculate Abnormal Return for the Horizon

The difference between the actual return and the expected return is the AR. It is reasonable to assume that the AR isolates the effect of the accident ("the event"). It is useful to examine how the ARs cumulate over time by calculating cumulative abnormal returns (CARs). For example, it can be of interest to examine how much return an investor would lose in the first week of the accident, the first month, the first quarter, and so on. Table 19.3 summarizes how ARs and CARs are calculated.

2.5 Step 4: Determine the Loss in Market Value due to Reputational Damage

ARs and CARs calculated in the previous step can be used to estimate a loss in the market value of a company during a day or a period of time, respectively. A loss in the market value is calculated from the abnormal rate of returns the company has experienced over a certain period and the market value of the company at the time of the accident. This loss

Table 19.2 Calculating the Returns Expected if There Were No Accident

Formula	Description
$r_{it} = \beta r_{mt} + \varepsilon_{it}$ $E[r_{it}] = \widehat{\beta} r_{mt}$	Estimated company beta (β) from the estimation window. Expected return on day t = the multiplication of the company beta to the actual market return on the day.

Table 19.3 Calculating Abnormal and Cumulative Abnormal Returns

Formula	Description
$AR_{it} = r_{it} - E[r_{it}]$ $CAR_{it} = \sum AR_{it}$	Abnormal return for the company on a particular day is the difference between the actual and the expected return for the day. Cumulative abnormal return for the period is the sum of the abnormal returns during the period.

Table 19.4 Estimating the Cost of Reputational Damage

Formula	Description
Loss in market value$_t$ = $AR_{it}(MV_{it*})$	The loss in the market value on day t is the abnormal return for the day multiplied by the market value of the company on the day of the accident, t^*.
$RL_{it} = AR_{it}(MV_{it*}) + AE_{it}$	Cost of reputational damage on day t is the sum of the loss in the market value on the day and the actual (or estimated) expense figure available on the day.

in the market value can then be compared with the actual expenses incurred through physical damage and the other direct expenses from the accident. Market efficiency posits that the change in the market value should correspond to the actual expenses incurred as a direct result of the accident. Any loss in the market value in excess of the actual (or estimated) expenses suggests the cost of reputational damage. The procedure for this step is summarized in Table 19.4.

3. APPLICATION OF THE APPROACH

On April 20, 2010, an explosion and subsequent fire occurred on the Deepwater Horizon semisubmersible mobile offshore drilling unit, which was drilling for BP in the Macondo Prospect, Gulf of Mexico. Control of the well was lost, resulting in a blowout—hydrocarbon gas and liquid rapidly and forcefully released from the well at rig level. The hydrocarbons found an ignition source on the drilling rig, and the resulting explosion caused the mobile offshore drilling unit to burn and sink. The impact of the Macondo incident was catastrophic: 11 workers died, 17 others were seriously injured, and 5 million barrels of oil reportedly spilled into the Gulf of Mexico, making it one of the largest environmental disasters in US history. The accident affected the lives of tens of thousands of people, from the families of those killed and injured to those whose livelihoods depend on the Gulf, as well as the broader oil and gas industry [1].

3.1 Net Worth Approach

The net worth approach can be used to estimate the cost of reputational damage to BP as a result of the accident. Eqs. (19.1) and (19.2) together state that the value of a company's reputation is estimated as the difference between market value of company's equity and net worth of the company. The loss in reputational damage can then be estimated as the change in the value of reputation after the accident as:

$$\text{Loss in reputational damage} = \text{value of reputation before the accident} - \text{value of reputation after the accident.} \quad (19.1)$$

$$\text{Value of reputation} = \text{market value of equity} - \text{book value of equity.} \quad (19.2)$$

The value of reputation for BP before the accident can be estimated using Eq. (19.2) as:

$$\text{Market value of equity before the accident} = \text{market price per share} \times \text{number of shares} = \$60.48 \times 18{,}732{,}459{,}000 = \$1{,}132{,}939.12 \text{ million}$$

Market price information was obtained from Yahoo finance [2]. Because the accident happened in the evening of April 20, 2010 [3], the market closing price on that day is used to calculate the pre-event market value. Information on the number of shares outstanding is obtained from the Annual Report 2009 of BP. Similarly, the market value of equity after the accident can be calculated as:

$$\text{Market value of equity after the accident} = \$60.09 \times 18{,}732{,}459{,}000 = \$1{,}125{,}633.46 \text{ million}$$

Because BP prepares annual reports once a year on December 31, book values of total assets and total liabilities of the company are assumed to remain the same around the period of the accident. In other words, in the absence of daily information on the book value of total assets and total liabilities of BP, the net worth of the company is assumed to have not changed after the accident. As a result, Eq. (19.1) is reduced to:

$$\text{Loss in reputational value} = \text{market value of equity before the accident} - \text{market value of equity after the accident} = \$1{,}132{,}939.12 \text{ million} - \$1{,}125{,}633.46 \text{ million} = \$7{,}305.66 \text{ million}$$

Therefore, on the first day after the accident, BP has lost more than $7 billion in reputational damage. One caveat to this estimate is that this loss measure does not take into account the loss in market value of equity due to the actual expenses incurred by the accident or any expenses incurred as a result of the accident. The estimate reflects only the loss in market value after the accident, whereas the loss could be reflective of both the loss in reputational damage and the loss in actual physical damage. As is common in such accidents, the estimated expenses to physical damages are not immediately available; thus, the immediate loss in market value may be regarded as the loss due to reputational damage.

As time passes and more up-to-date information on the company's assets and liabilities become available (eg, when the annual report for 2010 is released), the cost of reputational damage can be reestimated.

Using total assets and total liabilities information provided in Annual Report 2009 [4], the value of the reputation of BP as of December 31, 2009, can be estimated as:

$$\text{Market value of equity} = \$57.97 \times 18{,}732{,}459{,}000 = \$1{,}085{,}920.65 \text{ million}$$

$$\text{Net worth} = \text{total assets} - \text{total}$$
$$\text{liabilities} = \$235{,}968 \text{ million} - \$133{,}855 \text{ million} = \$102{,}113 \text{ million}$$

$$\text{Value of reputation on December 31,}$$
$$2009 = \$1{,}085{,}920.65 \text{ million} - \$102{,}113 \text{ million} = \$983{,}807.65 \text{ million}$$

These calculations can be repeated with information available on December 31, 2010 [5].

$$\text{Market value of equity} = \$44.17 \times 18{,}785{,}912{,}000 = \$829{,}773.73 \text{ million}$$

$$\text{Net worth} = \$272{,}262 \text{ million} - \$176{,}371 \text{ million} = \$95{,}891 \text{ million}$$

$$\text{Value of reputation on December 31,}$$
$$2010 = \$829{,}773.73 \text{ million} - \$95{,}891 \text{ million} = \$733{,}882.73 \text{ million}$$

Therefore, the cost of reputational damage for BP is:

$$\text{Change in value of reputation between 2010 and}$$
$$2009 = \$733{,}882.73 \text{ million} - \$983{,}807.65 \text{ million} = -\$249{,}924.92 \text{ million}$$

The company has lost \$249,924.92 million in reputational value from 2009 to 2010. A limitation to this estimation is that the actual expenses incurred in the accident are not accounted for. Thus, this approach may be applied when estimates of physical damage costs of the accident (and any other explicit costs) are not available or applicable. Another limitation is that the estimate is not dynamic because the value of net worth is available only when the company releases financial statements, which is once a year in the case of audited reports and quarterly for unaudited reports. Finally, the estimates are valid only for a short period after the date when the financial statements are released because the value of net worth is a flow variable that changes with time.

3.2 Event Study Approach

In this section, the event study approach is applied to estimate the cost of reputational damage from the accident.

To isolate the effect of the accident on the share price, the expected price dynamics, that is, expected returns on the share in the absence of an accident, is estimated first. The most common means of forming this expectation is by referring to historical price dynamics or returns. To get a reasonable representative of historical returns, a large number of historical returns is needed. It is common to use the past 250 trading days, that is, the past year, as a reasonable reference for historical returns. An advantage of using a long series of historical returns is that the series contains information about the influence that movements in the market can have on company returns. As a result, the movements in the share returns that are attributable to market movements, as indicated by company beta, can be estimated. This can be done in two phases.

First, the share covariability to the market returns, β, is estimated. Returns on a BP share in the 250 trading days before the accident are regressed on market returns. β can be expressed as

$$E[r_{it}] = 1.056r_{mt} + \varepsilon_{it}$$

where $E[r_{it}]$ is the return expected on BP shares on day t if there were no explosion and r_{mt} is the market return on day t. t-statistics for beta estimate is 17.79 and the regression R^2 is 0.557319.

Second, the beta estimate is used to predict expected returns in the period after the accident. Note that the expected returns obtained from the beta estimate are the expectations based on market movements, as suggested by the historical information. At this point it should be emphasized that there are other means to form expectations, such as the factor approach, as mentioned earlier. Yet the market model is one of the most-used approaches owing to its ease and intuitiveness. Despite being user-friendly, market model approach has its limitations. For instance, the value of beta estimated varies with frequency of data (ie, daily, monthly, or yearly) and the time period used in the estimation. However, the market model remains a popular choice among investment practitioners for predicting expected returns on a stock [6].

The beta estimate helps isolate the stock returns that are due to company-specific events, such as the accident. This can be done through establishing ARs. Table 19.5 shows the ARs in the 30 days after the accident.

For example, on the third day after the accident, a share of BP is expected to yield a return of:

$$\text{Expected return on April 23, 2010} = 1.056 \times 0.0071 = 0.0075$$

Actual returns on the day were 0.0055, which means that the AR for the day is −0.0020. In other words, given that the company has a beta of higher than 1, it is expected that the company returns would be more volatile than the market returns and that the company would experience a daily return of 0.75% on April 23 if there had been no accident. However, the company experienced a return far less than expected; the difference could be attributable to the accident's occurring on the evening of the 20th. As the news about the accident unfolded over the following days, the returns reacted accordingly. We can then estimate the CARs over a period of time by cumulating the ARs over time. For example, in the 5 days after the accident, BP shares experienced a cumulative loss of 5% over and above any variations caused by market movements. This is calculated as the sum of the ARs in the 5 days after the accident.

Fig. 19.2 shows the CARs in the 30 days after the accident. In the trading days after the accident, the company shares gradually lost their value; more than one

Table 19.5 Abnormal and Cumulative Abnormal Returns After the Accident

Date	Day(s) After Accident	Actual Return	Market Return	Abnormal Return	Cumulative Abnormal Return
21/04/2010	1	−0.0065	−0.0010	−0.0054	−0.0054
22/04/2010	2	−0.0090	0.0023	−0.0114	−0.0168
23/04/2010	3	0.0055	0.0071	−0.0020	−0.0188
26/04/2010	4	−0.0335	−0.0043	−0.0289	−0.0477
27/04/2010	5	−0.0277	−0.0237	−0.0027	−0.0504
28/04/2010	6	0.0178	0.0064	0.0110	−0.0394
29/04/2010	7	−0.0870	0.0129	−0.1006	−0.1400
30/04/2010	8	−0.0078	−0.0168	0.0099	−0.1301
03/05/2010	9	−0.0383	0.0130	−0.0521	−0.1822
04/05/2010	10	0.0199	−0.0241	0.0454	−0.1368
05/05/2010	11	−0.0041	−0.0066	0.0028	−0.1339
06/05/2010	12	−0.0130	−0.0329	0.0218	−0.1122
07/05/2010	13	−0.0256	−0.0154	−0.0093	−0.1215
10/05/2010	14	−0.0063	0.0430	−0.0518	−0.1733
11/05/2010	15	−0.0002	−0.0034	0.0034	−0.1699
12/05/2010	16	−0.0049	0.0136	−0.0193	−0.1892
13/05/2010	17	−0.0083	−0.0122	0.0046	−0.1846
14/05/2010	18	−0.0259	−0.0190	−0.0059	−0.1905
17/05/2010	19	−0.0064	0.0011	−0.0076	−0.1981
18/05/2010	20	−0.0259	−0.0143	−0.0108	−0.2088
⋮					
25/05/2010	25	0.0166	0.0004	0.0162	−0.2280
⋮					
01/06/2010	30	−0.1622	−0.0173	−0.1439	−0.3778

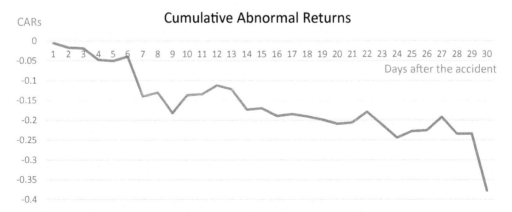

Figure 19.2 Cumulative abnormal returns in the 30 days after the accident.

third of the value was lost within the 30 trading days after the accident. This figure gives an indication of the extent to which BP shares lost their value, but they are not so informative about the reputational lost, because the loss in share value could be reflective of the physical losses incurred to the company, such as the cost of closing the leakage. Therefore, the next step is to compare the ARs with the physical expenses incurred.

The challenge in estimating the cost of reputational damage lies partly in estimating the cost of physical damages. In cases like BP, the physical costs manifest in many forms ranging from the costs incurred in sealing the leakage and cleaning the oil spilled onto the ocean to compensating the people and institutions affected directly and indirectly by the accident. Owing to the complexity of the case, it is not unusual for it to be years before a reasonable estimate of physical costs become available. At the time of this writing, that is, 15 years after the accident, the total physical costs of the accident are not yet established [7]. In Annual Report 2014, BP reports that the cumulative pretax income statement charges from the incident to the end of 2014 amounted to $43.5 billion. Market value of equity before the accident has been calculated earlier as $1,132,939.12 million. Thus, the cost of reputational damage in the 30 days after the accident can be estimated as in Table 19.6.

On the first day after the accident, the cost of reputational damage is:

$$RL_{day1} = -0.0054 \times \$1,132,939.12 \text{ million} + \$43,500 \text{ million} = \$37,382.13 \text{ million}$$

The positive value suggests that there is no cost due to reputational damage in the first day after the accident. This is because the abnormal loss in market value due to the accident is less than the actual physical costs of the accident. It is worth noting that the cost figure is from the annual report of the company 14 years after the accident. Because it is a retrospective figure and not the estimate available as of April 21, 2010, it appears as if there is no reputational loss on that day. This appears so for the next couple of days as well. From day 4 onward, the cost of reputational damage becomes apparent; it is estimated that the company within a week of the accident lost $13,545.51 million in market value in excess of the physical costs incurred and lost $384,484.87 million within 30 trading days of the accident.

Continuing the calculations for the rest of the year gives the estimated costs of reputational damage during the year after the accident. At the end of the year on December 31, 2010, the company is estimated to have lost $361,538.77 million in reputational damage from the accident. This estimate is more accurate than the estimate obtained from the net worth approach, which estimated the cost of reputational damage to be $622,996.11 million. Given that the net worth approach did not take into account the physical costs, we can see that the cost of reputational damage is overestimated in the net worth approach.

Table 19.6 Cost of Reputational Damage to British Petroleum in the 30 Days After the Accident

Date	Day(s) After Accident	Cost of Reputational Damage (in Millions of US dollars)
21/04/2010	1	37,390.35
22/04/2010	2	24,467.67
23/04/2010	3	22,236.02
26/04/2010	4	−10,512.12
27/04/2010	5	−13,545.51
28/04/2010	6	−1119.12
29/04/2010	7	−115,119.84
30/04/2010	8	−103,906.83
03/05/2010	9	−162,903.53
04/05/2010	10	−111,464.80
05/05/2010	11	−108,245.55
06/05/2010	12	−83,581.31
07/05/2010	13	−94,135.13
10/05/2010	14	−152,793.73
11/05/2010	15	−148,954.57
12/05/2010	16	−170,873.45
13/05/2010	17	−165,636.40
14/05/2010	18	−172,277.41
17/05/2010	19	−180,878.96
18/05/2010	20	−193,099.01
⋮		
25/05/2010	25	−214,787.19
⋮		
01/06/2010	30	−384,484.87
⋮		
31/12/2010	183	−361,538.77

4. CONCLUSIONS

This chapter illustrates a step-by-step tutorial on the application of the net worth approach and the event study approach to estimating reputational costs of a past accident. Despite the difficulty in estimating the value lost in reputational damage after an accident, the two approaches make it possible to assess the loss in reputational value. The net worth approach provides a quick approximation of the cost, whereas the event study approach gives a more rigorous figure as well as a possible range of costs.

REFERENCES

[1] US Chemical Safety and Hazard Investigation Board. Explosion and fire at the Macondo well — investigation report. Washington, DC, United States: Office of Congressional, Public, and Board Affairs; 2014.
[2] Yahoo. Yahoo finance. 2015. http://finance.yahoo.com/.

[3] BP. Deepwater Horizon accident and response. 2015. http://www.bp.com/en_us/bp-us/commitment-to-the-gulf-of-mexico/deepwater-horizon-accident.html.

[4] BP. Annual report and accounts. London, United Kingdom: BP plc; 2010.

[5] BP. Annual report and accounts. London, United Kingdom: BP plc; 2011.

[6] Bodie Z, Kane A, Marcus A. Investments. 8th ed. New York: McGraw Hill; 2009.

[7] BP. Annual report and accounts. London, United Kingdom: BP plc; 2015.

PART 4

Dynamic Risk Management

CHAPTER 20

Dynamic Risk Management in the Perspective of a Resilient System

T.O. Grøtan[1], N. Paltrinieri[1,2]
[1]SINTEF Technology and Society, Trondheim, Norway; [2]Norwegian University of Science and Technology (NTNU), Trondheim, Norway

1. INTRODUCTION

From the very beginning of the resilience engineering discourse in 2006, it has been questioned whether the concept is necessary [1]. From the variety of definitions, Roe and Schulman [2] even ask "what is *not* resilience?" Whatever the weaknesses and ambiguities of the definitions, resilience has become an important pivot point for safety and disaster management, but it is rarely mentioned in the risk parlance, except as a generic factor of mitigation.

Resilience implies a type of resistance that is metaphorically associated with organic flexibility. It is not a coincidence that butterflies' ability to coil up their long tubular mouthpart in a resting position is enabled by a rubbery protein named *resilin*, known for its ability to spring back into position [3]. The idea of resilient functioning may be synonymous with the popular idea of robustness. In that sense, it suggests a specific response related to a situation. However, on a generic level, system resilience does not entail clear starting or stopping points [4], but it is founded on dynamic conditions—especially when organizations experience their limits in anticipation, preparation, and preparedness. In fact, resilience refers also to inherently uncertain and fallible properties, more related to unexpected emergence than coherent mechanical dynamics.

In general, resilience is predominantly approached as (and thus hidden behind) a generic matter of complexity, ambiguity, and uncertainty about the future [5]. This is possibly due to risk assessment, management, and governance approaches, which are not acquainted with keeping pace with social dynamism and emergence, system discontinuities and ruptures, or organizational impermanence [6].

2. BACKGROUND: COMPLEXITY AND THE NOTION OF HIDDEN, DYNAMIC, AND EMERGENT RISKS

Our attention is specifically directed toward what we denote hidden, dynamic, and emergent (h/d/e) risks of complex systems. The notion of h/d/e is used to address risks that may be unknown in any sense—that is, risks that are ignored, forgotten, impossible

Dynamic Risk Analysis in the Chemical and Petroleum Industry
ISBN 978-0-12-803765-2

to identify, misunderstood or underestimated—stemming from dynamism and emergence and accommodating both ontological and epistemological uncertainty.

It is important to recognize that h/d/e risks are not exclusively external or exogenous to the system. Quite to the contrary, resilience implies that h/d/e risks are endogenous and relate to dynamics between the inside and the outside as well as systemic emergent properties.

Hence, h/d/e risks involve new challenges related to scientific knowledge, risk management methods, practical competence, regulation, and governance. These challenges are not confined to lack of certain knowledge available for a risk assessor, but they may relate to habituated structures of knowledge and uncertainty [7], such as the risk management vicious circles referred to in Section 3.1. Coherence may be only retrospectively visible. For this reason, it is important that the analyst does not to attempt to understand resilience implications in a passive way but tries to understand the understanding subject, that is, the resilient learning process itself.

3. STATE OF THE ART

3.1 Risk Management

Several examples of a framework addressing risk management or governance may be found in standards and related contributions in the literature, as follows:

- "Risk management: guideline for decision makers" by the Canadian Standard Association (standard CSA Q850-97) [8].
- "Risk management: principles and guidelines" by the International Organization for Standardization (standard ISO 31000:2009) [9].
- "Risk governance framework" by the International Risk Governance Council [10].
- "Risk and emergency preparedness assessment" by the Norwegian petroleum industry (standard NORSOK Z-013) [11].

The mentioned risk management frameworks unanimously address the following steps: preassessment, risk assessment, tolerability/acceptability judgment, risk management, and risk communication. Treatment of uncertainties is also emphasized, and different related practices are suggested. For instance, ISO 31000 defines risk as uncertainty to achieve an objective [9].

The International Risk Governance Council framework [10] distinguishes between uncertainty and ambiguity. Uncertainty refers to a lack of clarity in regard to the scientific or technical basis for decision-making, whereas ambiguity gives rise to several meaningful and legitimate interpretations of accepted risk assessments results. For instance, ambiguity may refer to potentially different values leading to a variety of interpretations.

Uncertainties can arise at different levels and moments of the risk management process and may be related to data, models, or the decision-making phase. Most of the risk management frameworks invite one to consider and acknowledge all forms of uncertainties, not only technical but also social. Decision-making under uncertainties usually relies on the consideration and comparison of multiple scenarios. However, continuous improvement is a fundamental element to treat uncertainties.

For instance, in the presence of emerging risks, differences among actual and expected results are likely to be present, owing to limits in experience and knowledge. The introduction of continuous improvement is fundamental to proceeding toward effective and efficient risk management.

Constant monitoring supports continuous improvement, which is already a recurrent step in the risk management frameworks presented. However, there are very few references in regard to what and how to monitor or measure. The monitoring process is often related to the level of achievement of objectives or to the adequacy of assumptions with observed consequences. On the contrary, the International Risk Governance Council framework insists on the monitoring of the following [10]:
- equity in repartition of risks and benefits among different categories of populations
- transparency and availability of information for various stakeholders

The dynamic risk management framework (DRMF) by Paltrinieri et al. [12] is presented as a development of the frameworks mentioned, focusing on continuous systematization of information from early signals of risk related to past events. This approach provides a support for the identification and assessment of potential atypical accident scenarios related to the substances, the equipment, and the site considered, capturing available early warnings or risk notions. Accident scenarios are defined as atypical if they deviate from normal expectations of unwanted events or worst-case reference scenarios [13].

DRMF, the schematization of which is shown in Fig. 20.1, has a characteristic shape that is open to the outside. This shows the need to avoid vicious circles and self-sustained processes and to open the process to external experience and early warnings. Following DRMF potentially allows integration of information about accident scenarios that we may not be aware that we do not know (unknown unknowns, *red line* (gray in print

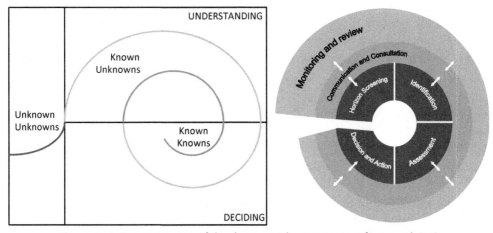

Figure 20.1 Representation of the dynamic risk management framework [12].

versions)). Increased awareness of h/d/e risks related to events we are now aware that we do not know (known unknowns, *yellow line* (light gray in print versions)) hypothetically leads to a learning and understanding phase (based on monitoring and review of accumulated information), supporting a following phase of decision on whether to assess such risks (based on communication and consultation). Horizon screening, hazard identification, assessment, and final decision/action are the steps needed by DRMF to thoroughly evaluate the risks associated with potential accident scenarios, which we are eventually aware we know (known knowns, *green line* (dark gray in print versions)) and can manage. There is no end to the process, but continuous reiteration to keep track of changes and process them for more effective and dynamic management of risk.

3.1.1 Risk and Complexity

The theory and practice of risk management has a proud heritage but is founded on less uncertain and ambiguous conditions than those implied by the radical h/d/e perspective. Owing to the increased (sociotechnical) complexity, lack of coherence, and speed of change in contemporary systems, the science and discipline of risk management continuously seeks to expand its relevance, reach, and grasp. The prevalent strategy is to incorporate uncertainty, ambiguity, and the knowledge dimension *per se* in the risk measure [14]. A key question is whether these attempts are effective for the radical h/d/e challenge, or if they actually produce a "rational disguise," inadvertently neglecting resilience. Do risk management practices contribute to the construction of a "rational facade" [15] for systems that are inherently more unstable than what is assumed?

Power [16] argues that professional and critical judgment of risk may be sacrificed for the sake of reputation (secondary risk), inciting an aversion against controversial or idiographic risk judgments. Paté-Cornell [17] argues that dominant metaphors within the risk management trade effectively function as excuses to wait for something to happen and for not acting on weak signals (eg, unexpected events that cannot be explained in standard ways, statistical outliers, singularities, idiographic experiences, and judgments). Regarding the potential implications of complexity, it is likely that risk considerations are also "trapped" in ways that incite reactiveness or deflection rather than proactivity in the face of complexity. Moreover, the actual impact of resilience, facilitated or necessitated by complexity, may be lost.

As previously mentioned, the black swan and the unknown/known concepts [13,18,19] have gained currency in the risk discourse. When we say "we do not know", who is "we"? It may be time to pay attention to the social construction dimensions of the risk knowledge utilized by risk assessors. The resilient subjects and their perspectives (the ones literally carrying the risks on their shoulders [20]) should be incorporated in the "we." However, the attempt to incorporate uncertainty and ambiguity reduction may come with a price: Stringency, clarity, and power of decision support for risk management may be lost in more uncertain and complex circumstances.

3.2 Resilience and Risk

The etymology of the resilience concept ranges over 2000 years [20], for the past 200 encompassing, for example, mechanics, child psychology, ecology, disaster risk reduction, and climate change adaptation. Disturbance, shock, complexity, emergence, and dynamism are the intrinsic premises for resilience in a wide sense. Continual performance variability due to intrinsic adaptations, easily ignored when "nothing" happens, is the norm rather than the exception. This is much in line with Karl Weick's characterization of high reliability organizations, for example, that "when nothing happens, a lot is happening" and "safety is a dynamic non-event" [6].

The potential scale of manifest change implied by the concept of resilience is wide. Generally speaking, it ranges between a "bounce back" from disturbance to a "nominal[1]" state, and a "bounce forward" to a new state of equilibrium, encompassing a fundamental change in underlying functioning [21]. In the former case (bounce back, Fig. 20.2), the change may be hardly noticeable but dynamic, temporary, and reversible. The key measure is time, process, and effort used to absorb the disturbance and return to normal. In the latter case (bounce forward, Fig. 20.3), the change is about dynamism, discontinuity, and emergence: a transformation process irreversibly turning into a new normal (state). The key measure is the degree of disturbance needed to exceed the ecological resilience and thus incite a flip between states. The terms *engineering resilience* and *ecological resilience* were coined by Gunderson and Holling [22], who place them within the generic[2] adaptive cycle shown in Fig. 20.4. This adaptive circle illustrates that "BF" is a third form of (transformational) resilience preceded by a saturated situation in which the range and capacity of both engineering and ecological is exceeded.

This third form of resilience (BF) corresponds to the notion of "socio-ecological" resilience as described by Comfort et al. [21], p19.

In the safety context, Hollnagel et al. [23] sparked the new and influential discourse denoted *resilience engineering*, still a pivotal point for safety science and practice in contemporary complex systems. Bergström et al. [24] analyzed the discursive patterns in a selection of the resilience engineering literature and grouped their findings along the axes of resilient *subject*, resilient *object*, and *rationale* for resilience. Resilient objects (the key phenomena studied) were mainly the link between resilience and successful (normal) operation, with the emphasis on dealing with disturbance or stress. Some of the literature also took a normative standpoint of resilience as being a good normative construct, inviting a problematic stance of potentially blaming victims for not being resilient (enough). Resilient subjects were either "sharp-end" operators, management, or teams or were attributed

---

[1] A given set of exceptional situations or conditions may be associated with the nominal. Hence, the change occurs when these are distorted or blurred.
[2] But ecologically inspired.

Figure 20.2 "Bounce back". *r*, engineering resilience. *(Adapted from Comfort LK, Boin A, Demchak CC. Designing resilience: preparing for extreme events. University of Pittsburgh Press; 2010.)*

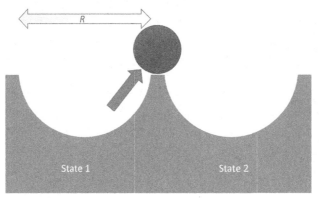

Figure 20.3 "Bounce forward". *R*, ecological resilience. *(Adapted from Comfort LK, Boin A, Demchak CC. Designing resilience: preparing for extreme events. University of Pittsburgh Press; 2010.)*

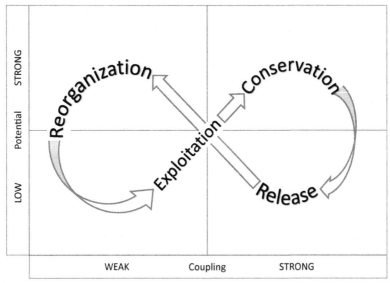

Figure 20.4 Adaptive circle. *(Adapted from Gunderson LH, Holling CS. Panarchy: understanding transformations in human and natural systems. Island Press; 2002.)*

to a depersonalized functional level of a system. The rationale for resilience varied between what is interpreted as two extremes: either as a strategy of precluding or outmaneuvering complexity based on preparedness or as a property that actually *thrives* on, and thus *seeks*, risk.

The resilience engineering literature is to some extent polemically flavored by the urge to demarcate a strictly rule-oriented safety approach from resilient functioning, its opposite. This sometimes takes the form of a safety 1 versus safety 2 (either/or) polemic, the latter representing the idealistic shape of a resilient system [25]. However, most companies and institutions do not enjoy the liberty to make an either-or choice, nor is it rational to do so from a safety point of view. Although the resilience literature does not preclude the existence and utility of procedures and rules, it is necessary to acknowledge more explicitly that resilience must unfold under not only the presence of rules but also the *imperative of compliance* with rule and procedures as the preferred and trusted way of instituting and communicating the intent and presence of safety [15]. This explicit acknowledgment has been shown to resonate with the pragmatic understanding of companies.

To proceed and make a difference with this pragmatic recognition, it is not enough to focus solely on one type of resilient subject [26]. For this to work, it is necessary to focus both on the necessary *margins of maneuver experienced* by the operational subjects and on the *space of maneuver mandated* by the managerial subjects according to a risk-oriented rationale and to reconcile them continually according to shifting conditions and experiences. This approach is commensurate with the identified need to agree on dynamical "bandwidths" in processes of high reliability management [2].

The compliance versus resilience (CvR) relationship [15] has deeper implications than the occasional combination of two different safety principles put on a par. Aiming for resilience implies an urge for and appreciation of emergent "salutogenic" processes (focusing on factors that support system health rather than disease) with no clear starting or stopping points [4]. However, Comfort et al. [21] affirm that resilience is not necessarily "a nice place to be" for human actors, raising the awareness that resilience is a fallible process and that adaptive traps are lurking [27] at every corner. Hence, the popular notion of resilience as being "prepared to be surprised" can be somewhat deceiving.

The term resilient organizations may interfere with other notions, such as high reliability organizations or high reliability management. The conceptual nuances are addressed elsewhere [2]. A key difference may be that some organizations labeled resilient may afford to fail and learn, whereas others may not: they are fault *intolerant* with respect to some specific outcomes. However, engineering as well as ecological resilience properties might still be a meaningful option, so long as the disturbances do not affect the critical outcomes.

Any assessment of resilience, and especially the risk implied by its presence, is a moving target, always embedded in epistemic uncertainty. Any retrospective coherence drawn from a complex system, with or without resilient properties, may be an anecdotal fallacy. We advance on the presumption that resilience implies a potential change of operating conditions and characteristics of a system, which may have an impact on risk. This impact should be considered if the risk is assessed on prior presumptions from normal or nominal[3] working conditions.

The possibility for recognition and assessment of such an impact is the theme of this chapter. The quest for understanding of risk variations under the presence of resilience is initially hypothesized as a search for available "decisive occasions" for understanding resilience. A preferred strategy for capturing decisive occasions is based on continually discovering brittleness and identifying resilient practices, when organizations experience their limits in anticipation, preparation, and preparedness. We need to identify such occasions and devise how to make sense of them with respect to risk.

4. THE SoS PROXY APPROACH

What is the risk contribution from the presumed or observed presence of resilience? We need to define (1) when we can see manifestations of resilience and (2) how we can assess their impact on risk.

The first point should be addressed by management, asking how to sensitize the risk management process to the changing system characteristics owing to resilient functioning, before assessing the implications of resilient functioning regarding risk (second point).

These considerations will not be straightforward. For example, an observation of a successful resilient episode could have various implications for the future, for example:

- a presumed positive effect in terms of (anecdotal) evidence of enhanced processes of preclusion, mitigation, or recovery
- a presumed negative effect in terms of amplified damage when eventually failing from higher grounds, risk compensation behavior, or higher propensity to seek borderline conditions

The scope of assessment is not necessarily on discrete events. It might be asked whether a series of successes has similar effects. Even the opposite (series of failures) may signify a turning point due to accumulated learning.

4.1 Sensitization to New Conditions: How to Recognize Resilient Functioning That Might Affect System Characteristics

In our view, resilient episodes cannot be understood out of context. The CvR model [15] (Fig. 20.5) is used to structure the argument. Resilience is positioned in context

---

[3] A given set of exceptional conditions may be associated with nominal status. Change occurs when these conditions are distorted or blurred.

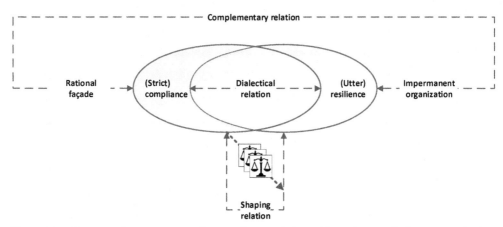

Figure 20.5 The compliance versus resilience relations. (*Adapted from Grøtan TO. Organizing, thinking and acting resiliently under the imperative of compliance. On the potential impact of resilience thinking on safety management and risk consideration. Trondheim: NTNU; 2015.*)

of compliance, in an asymmetrical but *complementary* relation. Such a relation is *dialectical* (examining the opposing ideas of absence versus presence) and mutually *shaping*, from which a successive series of CvR reconciliations emerge.

A time series of CvR reconciliations constitutes a dynamic SoS, in principle adapted to the changing circumstances presumed and experienced in an informed safety management process.

A model is needed for the safety management process to identify and grasp the occasions that can be used to sensitize an iteration of the CvR relation. For that purpose, a drift model as in Fig. 20.6 is applicable. The model is based on Snook's concept of practical drift [29]. The model has been modified to illustrate that this drift is not necessarily a "drift into failure" but might as well be a "drift into success" in a complex environment as

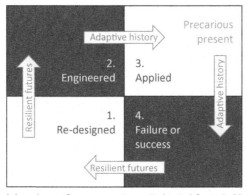

Figure 20.6 The drift model and significant occasions. (*Adapted from Dekker S. Drift into failure: from hunting broken components to understanding complex systems. Ashgate Publishing Limited; 2012.*)

devised by Dekker [28] and thus a manifestation of resilience as a positive outcome of complex system properties.

The drift metaphor is recurrent and recursive in the sense that, for example, technical revisions and redesigns, organizational changes, failures, incidents, accidents, recoveries, and not least mastery of unexpected situations may represent decisive occasions in terms of manifestations or potential restarts of drift at different scales. A vigilant organization will not run out of decisive occasions inviting sense-making work. Three types of occasions are available:

- *Adaptive history*: the crucial question is "how have we become what we are?" Does this invite a revision of the prevailing SoS? Is the resilience part of the CvR relation overrated or underrated?
- *Precarious present*: the crucial question is "where are we now?" What is at stake? Where is the expected brittleness positioned? How can it be characterized? Is it commensurate with the present SoS?
- *Resilient future*: the crucial question is "what do we want to become?" Any SoS is possible, but the credibility of the choice depends on the "we." Hence, the operational staff must be involved in this projection. However, any chosen answer should be justified by critically devising the future capabilities, based on rated merit or on substantial preparation.

4.2 How Do We Derive Risk-relevant Knowledge from Resilient Functioning?

The remaining challenge is to derive risk-related knowledge from the SoS and the drift model, which here is a proxy concept for the real risk that is not available. However, the SoS proxy allows for reorienteering to keep pace.

As represented in Fig. 20.7A, this can be done in a pulsed manner [15], in which the pulse beat is driven by the occasions derived from the drift model. For each pulse beat, there is an expansion phase, a contraction phase, and a succeeding blood (that is, information) flow that lasts until the next beat. Hence, the pulse of risk (PoR) is a correlate [15] of the SoS as follows:

- In the *expansion* phase, the current CvR reconciliation is critically examined.
- In the *contraction* phase, any SoS change is followed by (1) a direct revision of existing risk assessments (a scheme for reassessment of risk based on complexity forms is offered by Grøtan [30]), and (2) an identification of a need for reorientation of the risk horizon, that is, the information outlook that corresponds to the revised SoS.
- In the *flow* phase, organizational attention is reorganizing according to the new risk horizon derived from the SoS.

This PoR approach [15] can incorporate and benefit from other approaches, for example, the DRMF approach represented in Fig. 20.1, which is a systematic attempt to reduce uncertainty under specific conditions. As illustrated in Fig. 20.7A and B, the

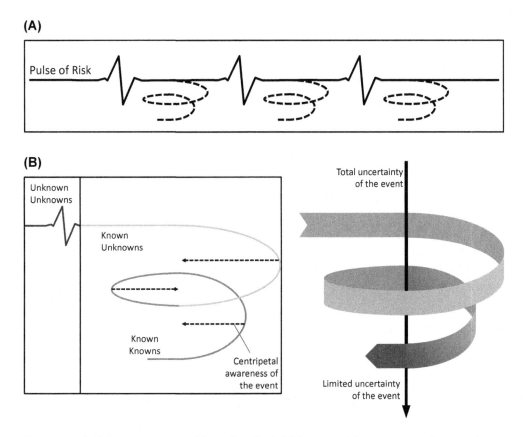

Figure 20.7 (A,B) Representations of the pulse of risk. (C) Dynamic risk management framework reorientation triggered by PoR.

PoR approach can be used to successively reorient and reinitialize the DRMF process. PoR allows for a shift in the DRMF perspective: from a two-dimension process (see Fig. 20.1) designed to continuously integrate exogenous information into risk evaluation to a three-dimension process (see Fig. 20.7B) convolving around the knowledge axis introduced by Aven [14] and iterated on the endogenous conditions provided by PoR.

Iteration of the DRMF approach after a PoR beat is advisable (Fig. 20.7A). New awareness of potential accident scenarios may be raised as a result of the reorganized organizational attention (Fig. 20.7B). Unknown unknowns may turn into events that we are aware we do not know (known unknowns). Such development activates the risk management process introduced in Section 3.1, which represents the application of the dynamic risk analysis methods described in the previous chapters. The objective of the DRMF process is assessing and metabolizing information on potential accident scenarios to continuously improve the current risk picture and limit uncertainties in the management of such risk. Decreasing uncertainty would allow the definition of such scenarios as known knowns. The force driving this process is the awareness of the potential accident event, represented as a radial centripetal force in the DRMF spiral movement (Fig. 20.7B).

5. CONCLUSIONS

Resilience impact on risk is addressed as a matter of understanding the specific changes in operating conditions. An episode of successful adaptation is not a guarantee for the future. An adaptive failure may provide the grounds for increased resilience by virtue of learning. Risk contribution from presumed or observed resilience can be addressed by various strategies. In particular, this chapter has presented a proxy approach aiming to recognize resilient functioning affecting system characteristics. Under new conditions, an organization may reorienteer in a pulsed manner, triggering iteration of the DRMF approach. DRMF has the potential to update the overall risk picture by deriving risk-relevant knowledge from resilient functioning. DRMF summarizes the dynamic risk analysis methodologies described in this entire book, and, in this chapter, it is addressed from the perspective of an impermanent, organizational, and resilient system.

REFERENCES

[1]   Hale A, Heijer T. Is resilience really necessary? The case of railways. In: Woods PDD, Leveson PN, Hollnagel PE, editors. Resilience engineering: concepts and precepts. Ashgate Publishing Limited; 2012.
[2]   Roe E,M, Schulman P. High reliability management: operating on the edge. Stanford (CA, USA): Stanford Business School Books; 2008.
[3]   Neal J. Living with insects. 2016.
[4]   Cavallo A, Ireland V. Preparing for complex interdependent risks: a system of systems approach to building disaster resilience. Int J Disaster Risk Reduct 2014;9:181−93.
[5]   IRGC. The emergence of risks: contributing factors. Geneva (Switzerland): International Risk Governance Council; 2010.
[6]   Weick K. Making sense of the organization: volume two: the impermanent organization. Hoboken (New Jersey, USA): John Wiley & Sons; 2009.
[7]   Berger PL, Luckmann T. The social construction of reality; a treatise in the sociology of knowledge. Garden City (NY): Doubleday; 1966.
[8]   CSA. Risk management: guideline for decision makers. Standard Q850−97. Ottawa (Canada): Canadian Standard Association; 2007.
[9]   ISO. Risk management — principles and guidelines. ISO 31000:2009. Geneva (Switzerland): International Organization for Standardization; 2009.
[10] IRGC. Risk governance deficits. Geneva (Switzerland): International Risk Governance Council; 2009.
[11] NORSOK. Risk and emergency preparedness assessment. Z-013. Oslo (Norway): Standards Norway; 2010.
[12] Paltrinieri N, Khan F, Amyotte P, Cozzani V. Dynamic approach to risk management: application to the Hoeganaes metal dust accidents. Process Safety and Environmental Protection 2014;92:669−79.
[13] Paltrinieri N, Dechy N, Salzano E, Wardman M, Cozzani V. Lessons learned from Toulouse and Buncefield disasters: from risk analysis failures to the identification of atypical scenarios through a better knowledge management. Risk Analysis 2012;32:1404−19.
[14] Aven T. Practical implications of the new risk perspectives. Reliability Engineering & System Safety 2013;115:136−45.

[15] Grøtan TO. Organizing, thinking and acting resiliently under the imperative of compliance. On the potential impact of resilience thinking on safety management and risk consideration. Trondheim: NTNU; 2015.

[16] Power M. The risk management of everything: rethinking the politics of uncertainty. Demos; 2004.

[17] Paté-Cornell E. On "Black Swans" and "Perfect Storms": risk analysis and management when statistics are not enough. Risk Analysis 2012;32:1823—33.

[18] Aven T. Implications of black swans to the foundations and practice of risk assessment and management. Reliability Engineering & System Safety 2015;134:83—91.

[19] Haugen S, Vinnem JE. Perspectives on risk and the unforeseen. Reliability Engineering & System Safety 2015;137:1—5.

[20] Perin C. Shouldering risks: the culture of control in the nuclear power industry. Princeton: Princeton University Press; 2005.

[21] Comfort LK, Boin A, Demchak CC. Designing resilience: preparing for extreme events. University of Pittsburgh Press; 2010.

[22] Gunderson LH, Holling CS. Panarchy: understanding transformations in human and natural systems. Island Press; 2002.

[23] Hollnagel E, Woods DD, Leveson N. Resilience engineering: concepts and precepts. Ashgate; 2006.

[24] Bergström J, van Winsen R, Henriqson E. On the rationale of resilience in the domain of safety: a literature review. Reliability Engineering & System Safety 2015;141:131—41.

[25] Hollnagel E. Safety-I and safety-II: the past and future of safety management. Ashgate Publishing Company; 2014.

[26] Grøtan TO, van der Vorm J, Macchi L. Training for operational resilience capabilities (TORC): 1st concept elaboration. Trondheim (Norway): SINTEF; 2015.

[27] Woods DD, Branlat M. Basic patterns in how adaptive systems fail. In: Hollnagel E, Pariès J, Woods DD, Wreathall J, editors. Resilience engineering in practice. Farnham (UK): Ashgate; 2011.

[28] Dekker S. Drift into failure: from hunting broken components to understanding complex systems. Ashgate Publishing Limited; 2012.

[29] Snook SA. Friendly fire: the accidental shootdown of U.S. Black Hawks over Northern Iraq. Princeton, Chichester (NJ): Princeton University Press; 2000.

[30] Grøtan TO. Assessing risks in systems operating in complex and dynamic environments. In: Albrechtsen E, Besnard D, editors. Oil and gas, technology and humans: assessing the human factors of technological change. Farnham: Ashgate; 2013.

Conclusions

Risk analysis in the chemical and petroleum industry is evolving. New dynamic techniques have evolved following recent methodological developments. The concept of dynamicity has gone beyond time dependence and online monitoring; it now encompasses progressive calibration and refinement of nonlinear repetitive processes and reacting and adapting to changes and new information flows. Despite innovative and effective approaches, industry has shown little interest in implementation of these concepts. This book explains and discusses several solutions for dynamic risk analysis and integration through parallel disciplines to provide concrete support to industrial application.

In Part 1, the introduction of Chapter 1 offers an overview on recent progress and applications of industrial risk analysis. Accomplishments and limitations are pointed out, but special attention is dedicated to the inability of risk analysis to update the risk picture. This led to the development of recent dynamic risk assessment approaches aiming to provide real-time support to decision-making. Chapter 2 reports an overview of definitions outlining low-probability high-impact events, classification of which is particularly challenging because of their rarity. To a certain extent, these definitions lead to adoption of continuously improving models and classifications to keep track of the ever-changing industrial environment.

Part 2 focuses on the core of risk analysis, and subsection 2.1 addresses the step of hazard identification. Chapter 3 introduces a specific dynamic methodology for this step. The method is conceived to provide comprehensive hazard identification of industrial systems, joined to a process of continuous improvement of the assessment results. Such improvement is possible by means of systematization of information from early risk signals. A step-by-step tutorial for the application of this technique is provided in Chapter 4. A representative application is also carried out on liquefied natural gas regasification technologies, when related risks can be well known to academics and experts, but safety professionals may disregard limited-experience scenarios.

Analysis of initiating events has an important role in dynamic risk analysis and is discussed in subsection 2.2. Chapter 5 presents reactive approaches of probability updating that adapt to stochastic events and processes over time. Such approaches may be represented by Bayesian methods, for example, hierarchical Bayesian analysis and the Bayesian network, and the chapter describes the state of the art of their application.

Chapter 6 shows how collection and monitoring of specific indicators may enable dynamic and proactive risk assessment. Proactive approaches may be grouped into four classes, and the technique introduced in this chapter (the risk barometer) is a representative example of one of them. Such a technique pays particular attention to human and organizational factors and the capability to support key safety decision-makers in short- and medium-term planning.

Dynamic Risk Analysis in the Chemical and Petroleum Industry
ISBN 978-0-12-803765-2

Tutorials and application examples of Bayesian inference-based dynamic risk assessment and the risk barometer are described in Chapter 7. The methods are applied to the same representative case to highlight similarities and differences and to provide support for the selection of methods. Chapter 8 defines in detail benefits and limitations of reactive and proactive approaches. Moreover, potential complementarities are identified to pave the way for further improvement and mutual integration of the two methodologies.

Analysis of consequences is the main topic of subsection 2.3. Computational fluid dynamic models are treated in Chapter 9 because they have the potential to overcome simplified tools and conservative assumptions for consequences assessment. Such approaches aim to capture the interaction evolution of the released hazardous substances and the environment geometry. The necessary steps for carrying out a large-scale dispersion study by means of computational fluid dynamic models are described in Chapter 10. Moreover, the methodology is applied in the analysis of the accident in Viareggio (Italy) in 2009: a flash fire that occurred after the derailment of a freight train.

The existing methodologies for severity assessment of runaway reactions are based on standard experimental activities, which may be relatively time consuming and economically unfeasible. For this reason, Chapter 11 describes related fundamental concepts and introduces dynamic risk assessment based on a minimum set of data. Chapter 12 elaborates on this approach by presenting a methodology tutorial and showing analysis results of the thermal decomposition of an organic peroxide. The analysis is based on an adiabatic experiment, providing data that can closely predict large-scale behavior.

Chapter 13 addresses the state of the art of traditional risk metrics and the related limitations. Dynamic risk visualization solutions are also suggested to overcome such limitations and offer not only graphical user interfaces but also a way to communicate a continuously updating risk picture.

Several parallel disciplines interact with risk analysis to provide sound details and improve the assessment process. Such interplay is treated in Part 3. Human errors are discussed in subsection 3.1, which describes representative human reliability analysis techniques and related differences and complementarities for potential industrial applications (Chapter 14). Fundamental importance is acknowledged for human factors contributing to major oil and gas accidents. For this reason, Chapter 15 provides an application tutorial of a human reliability analysis technique (standardized plant analysis risk) used for the petroleum sector and an example of application: the analysis of a drive-off scenario of a semisubmersible drilling unit.

Subsection 3.2 deals with costs and benefits of safety. Chapter 16 describes cost-benefit analysis in the evaluation of investments in safety measures under an economics perspective. Cost-benefit analysis may improve decision-making because it allows prioritizing and refocusing on safety rather than on short-term organizational objectives. A disproportion factor may be used as an adjustment in favor of safety when partial quantitative data are available. A multicriteria decision-making approach is also suggested in case of lack of data. Both the methods are described in Chapter 17 by means of tutorials and specific examples of application.

To carry out a comprehensive cost-benefit analysis, the potential damage to the reputation of a company subject to a major accident should be assessed based on related experience. Subsection 3.3 addresses this specific aspect. In particular, Chapter 18 introduces emerging approaches to quantifying the reputational damage of past events experienced by companies. Advantages and disadvantages of the approaches are also discussed. Chapter 19 provides a step-by-step tutorial for the application of event study methodology. An example is also carried out on the explosion on April 20, 2010, at the Deepwater Horizon oil rig operated by British Petroleum.

Dynamic risk management is the object of the last part of the book (Part 4). Chapter 20 discusses this in the perspective of a resilient system, where resilience is about dealing with the unexpected and the unprecedented. Various strategies are explored in this chapter, with special emphasis on a proxy approach as a means of sensitizing risk assessments to the implications of complexity and traces of presumed resilience.

As evident from this summary, the main conclusion to draw is that relevant recent research has provided a framework for approaching dynamic risk assessment and supporting its management by the organization. Within this framework, the book addresses the several factors constituting a risk picture and describes specific methods to allow for progressive refinement of results. In this way, the risk picture is continuously improved and adapted to reflect actual system conditions. However, such methodological achievements need to be consolidated to be attractive for implementation. No best practices or standards have been available until now to support dynamic risk assessment. It is hoped that this book will pave the way for a process of acceptance and standardization, contributing to the enhancement of safety and sustainability in the chemical and petroleum industry.

INDEX